Smart Transportation
AI Enabled Mobility and Autonomous Driving

Editors

Guido Dartmann
Group Leader – Distributed Systems
Trier University of Applied Sciences
Hoppstädten-Weiersbach, Germany

Anke Schmeink
Professor and Group Leader – Information Theory
and Systematic Design of Communication Systems
RWTH Aachen University, Aachen, Germany

Volker Lücken

Houbing Song
Assistant Professor
Embry-Riddle Aeronautical University
Daytona Beach, Florida, USA

Martina Ziefle
Full Professor of Communication Science
RWTH Aachen University, Aachen, Germany

Giovanni Prestifilippo
Managing Director
PSI Logistics GmbH, Berlin, Germany

CRC Press
Taylor & Francis Group
Boca Raton   London   New York

CRC Press is an imprint of the
Taylor & Francis Group, an **informa** business

A SCIENCE PUBLISHERS BOOK

First edition published 2021
by CRC Press
6000 Broken Sound Parkway NW, Suite 300, Boca Raton, FL 33487-2742

and by CRC Press
2 Park Square, Milton Park, Abingdon, Oxon, OX14 4RN

Library of Congress Cataloging-in-Publication Data

Names: Dartmann, Guido, 1980- editor.
Title: Smart transportation : AI enabled mobility and autonomous driving / editors, Guido Dartmann, Anke Schmeink, Volker Lücken, Houbing Song, Martina Ziefle, Giovanni Prestifilippo.
Description: First edition. | Boca Raton, FL : CRC Press, 2021. | Includes bibliographical references and index. | Summary: "The book provides a broad overview of the challenges and recent developments in the field of smart mobility and transportation. This includes technical, algorithmic as well as social aspects of smart mobility and transportation. It starts with new ideas for services and platforms for future mobility. Then new concepts of artificial intelligence and the implementation in new hardware architecture are discussed. In the context of artificial intelligence, new challenges of machine learning for autonomous vehicles and fleets are investigated. Finally, the book concludes with human factors and social questions of future mobility concepts. The goal of this book is to provide a holistic approach towards smart transportation. The book reviews new technologies cloud, machine learning and communication for fully atomized transport, catering to the needs of citizens. This will lead to complete change of concepts in transportation"-- Provided by publisher.
Identifiers: LCCN 2021014170 | ISBN 9780367352967 (hardcover)
Subjects: LCSH: Transportation--Technological innovations. | Intelligent transportation systems. | Automated vehicles. | Communication and traffic.
Classification: LCC HE151 .S555 2021 | DDC 629.04/6--dc23
LC record available at https://lccn.loc.gov/2021014170

ISBN: 978-0-367-35296-7 (hbk)
ISBN: 978-1-032-02883-5 (pbk)
ISBN: 978-0-367-80815-0 (ebk)

Typeset in Times New Roman
by Innovative Processors

Preface

Every day, billions of people commute to work, drive to shopping locations or travel to visit friends. As society, mobility requirements, energy needs and technology options are currently undergoing a fundamental (change,) as (the demands) for a clean, sustainable, efficient and safe mobility are coming into effect. This will lead to a complete change in the latest transportation concepts.

The aim of the book is to present a holistic approach to smart transportation. The contributions are dedicated to new cloud, machine learning and communication technologies for a fully atomized transport, taking into account the needs of citizens. The book offers readers a broad overview of challenges and recent developments in the field of smart mobility and transportation.

This includes both technical and social aspects in order to cover different perspectives on this topic. The book begins with recent concepts for services and platforms for future mobility.

Subsequently, novel concepts of artificial intelligence and their implementation in autonomous vehicles are discussed. In connection with artificial intelligence, new challenges of machine learning and algorithms for autonomous vehicles and fleets are also investigated. In this regard, an insight is also given into the changing way of interactions between different stakeholders, leading to a joint and agile development of the technology component. Beyond the technical and infrastructural issues of autonomous driving, the book also addresses the role of human factors and social questions of future mobility concepts.

The book is organized in 11 chapters which are clustered in three sections:

A first section focusses on large-scale perspectives of autonomous driving. In this first part, we present five chapters that deal with a large-scale view on autonomous mobility. We start with a chapter on the social acceptance of autonomous driving and discuss the role and inclusion of users. In Chapter 2, we present the idea of a best-practice project in the city of Aachen, where we additionally present a large-scale simulation of a mobility-on-demand system for the same reference city in the 3rd chapter. Chapter 4 presents latest results of optimization techniques for on-demand-systems for large fleets of autonomous vehicles. Finally, Chapter 5 concludes this part with innovation management and cooperation concepts for an agile future transformation of transportation. A second section addresses secure communication of future smart mobility. Here, chapters discuss the communication aspects of future systems for autonomous mobility. Chapter 6 discusses new distributed ledger techniques and Chapter 7 presents a pseudonym management by blockchain, which can guarantee secure transactions in the system. Finally, the third part explores autonomous driving techniques, including simulation platforms, deep and reinforcement learning and functional safety. Chapter 8 presents an overview of the latest simulation platforms for autonomous driving. Chapter 9 introduces recent deep learning techniques of visual computing for autonomous driving. These results are extended to reinforcement learning and unsupervised learning techniques in Chapter 10. Finally, Chapter 11 concludes the book with latest results on the functional safety of deep learning techniques on autonomous driving.

In the following, the different chapters are listed.

Part 1: Larges-scale views on autonomous driving: social aspects, best-practice projects, simulations and optimization of mobility-on-demand systems:

(1) Social Acceptance of Autonomous Driving - the Importance of Public Discourse and Citizen Participation (Brell et. al)

(2) Artificial Intelligence for Autonomous Vehicle Fleets: Desired Requirements, Solutions and a Best Practice Project (Prestifilippo)

(3) Large-scale Simulation of a Mobility-On-Demand System for the German City Aachen (Matthias et. al)

(4) Artificial Intelligence for Fleets of Autonomous Vehicles: Desired Requirements and Solution Approaches (Dziubany et al.)

(5) Future Urban Mobility: Designing New Mobility Technologies in Open Innovation Networks (Hedderich et al.)

Part 2: Secure communication of future smart mobility:

(1) Communication and Service Aspects of Smart Mobility: Improving Security, Privacy and Efficiency of Mobility Services by Utilizing Distributed Ledger Technology (Ayad et. al)

(2) Pseudonym Management through Blockchain: Cost-Efficient Privacy Preservation on Smart Transportation (Bao et. al)

Part 3: Autonomous driving: Simulation platforms, deep and reinforcement learning and functional safety:

(1) Simulation Platforms for Autonomous Driving and Smart Mobility: Simulation Platforms, Concepts, Software, APIs (Creutz et al.)

(2) Deep-Learning Based Depth Completion for Autonomous Driving Applications (Lanius et. al)

(3) Artificial Intelligence Based on Modular Reinforcement Learning and Unsupervised Learning (Mohammed et. al)

(4) Functional Safety of Deep Learning Techniques in Autonomous Driving Systems (Chen et al.)

Acknowledgements

We would like to thank all people contributed to our book project over the last years 12 months. Especially we would like to thank all authors of the chapters and all reviewers helping to improve the quality of each chapter. Finally, we would like to thank CRC for their fast and professional support during our editorial work.

This work has been partly (chapters 1, 2, 3, 4, 8 and 10) funded by the Federal Ministry of Transport and Digital Infrastructure (BMVI) within the funding guideline "Automated and Connected Driving" under the grant number 16AVF2134.

Guido Dartmann
Anke Schmeink
Volker Lücken
Houbing Song
Martina Ziefle
Giovanni Prestifilippo

Contents

Social Acceptance of Autonomous Driving – The Importance of Public Discourse and Citizen Participation

Teresa Brell[1], Ralf Philipsen[2], Hannah Biermann[2] and Martina Ziefle[2]*

[1] umlaut Solutions, Aachen, Germany
[2] Chair of Communication Science, Human-Computer Interaction Center,
RWTH Aachen University, Germany

1.1 Introduction – Automated Driving: Aims, Status Quo, Challenges

The use of autonomous vehicles for automated city centers enables new, innovative mobility solutions, which combine elements of private and public local passenger transport and enables users to achieve a high degree of mobility. In order to fully exploit the potential of autonomous vehicles for sustainable urban mobility and for a successful and sustainable rollout it is important to understand the requirements of new mobility solutions, their economic, legal and social challenges and also to reflect on which contexts the implementation of a new technology is seen as a threat or perceived as an enhancement.

Mobility is in a state of constant change because the technical possibilities are increasing and the fast paced development of technology in the mobility sector, especially in road traffic, opens up new ways to travel. These possibilities can bring both benefits and risks. On the one hand traffic infrastructure, vehicles and pedestrians offer new possibilities to intervene in and observe critical situations through their connection via driver assistance systems [1]. At the status quo today, connectivity supports a way to significantly improve road safety by integrating all traffic participants [2]. Further, users understand that automated driving increases safety in traffic. Automatic driving functions for example allow cognitive relief on longer journeys, which is particularly useful for older drivers [3, 4, 5].

*Corresponding author: ziefle@comm.rwth.-aachen.de

On the other hand, the possibility of gaining valuable data and information may encourage the ill-intended tracking of users and this collected shared information might lead to concerns from potential users. The possible loss of control may result in serious privacy and trust issues [5, 6]. Additionally, uncertainty is fueled by negative press, e.g. the first Tesla crash in 2016 and the latest in early 2019, which caused great insecurity among drivers and led to questions about the dangers and consequences of autonomous driving [7, 8].

Thus, the innovation management of introducing new technologies is often a lengthy process, accompanied by social and societal skepticism [9, 10]. This is particularly true in the case of complex and fundamentally new technology developments, which, from a societal point of view, may have undeniable advantages, but at the same time trigger basic fears among users, e.g. loss of control, outside interference and loss of responsibility [11, 12], as it is in the case of automated or autonomous driving. For long experienced (professional) drivers, autonomous vehicle technologies might also be seen as a violation of tradition, because for the first time, the car is in control and not the driver [13], who experiences a sense of loss of control and has a fear of being controlled by the technology [14, 15]. This is an important and critical point since studies have shown that rearranging the control might be a necessary and important step for the reduction of road accidents, because the main reason for accidents in traffic is the human driver [16]. Hence, trust in and acceptance of V2X-technology and autonomous driving is a significant factor for a successful rollout of autonomous driving though insufficiently explored so far [17, 18]. In order to address possible acceptance problems in a systematic way and to communicate acceptance-relevant topics in time, it is necessary to identify them first [19].

But fundamental changes are not possible without recognizing the needs, emotions and requirements of the targeted users, and without including them in the transformation process [20, 21]. The transition to autonomous traffic will also be a challenging process: Driving a car is perceived as a deeply rooted part of social life (at least in the western world) and definitely enhances perceived social esteem [22, 23]. That is why introducing autonomous driving into road traffic is a sensitive and at the same time exciting step. Driving has an emotional value, which is why autonomous driving could be seen as a violation of tradition: the vehicle (or the infrastructure) would control the ride instead of the driver, who now has the role of a passenger. Not only the feeling of loss of control but also the fear of being controlled by the technology may come up [24].

Not only in the context of autonomous driving and the required infrastructure but also in other contexts (such as energy infrastructure) the involvement of users is an important and impactful choice. With the turn towards many large-scale projects in the context of e.g. renewable energies and novel mobility infrastructure, social acceptance of the infrastructure needed to provide successful rollouts of technological infrastructure in urban environments evolved as a major topic within a large strand of research.

Evidently, without adequate understanding and support by the communities, the cities and the public, it will be difficult if not impossible to meet the societal challenges of energy needs as well as the changes in the mobility sector and to put the necessary infrastructure into practice [25, 27, 27].

Taking all given information into account, the current work sets two focuses: One is to stress the importance of social acceptance for the successful innovation management of autonomous vehicles. The second focus is to illustrate one specific measure to show a practical example of citizen's science in Aachen: The citizens' dialogue of the project APEROL.

This work concludes with the generalizability and transferability of the procedure and the results to other mobility areas.

1.2 Social Acceptance – Its Role for and the Inclusion of Potential Users

Social acceptance focuses on the approval, positive reception, and sustainable implementation of (novel) technologies [28]. Within academic research, thus, individual usage motives, but also possible barriers are explored in order to shed light on the attitudes and public opinions towards the technology. Research revealed that the perceived risk of a novel technology and the probability of non-acceptance or even protest are inversely related to the familiarity and the prevailing information depth about a certain technology [15, 18]. In addition, there is evidence that personality factors (risk avoidance, innovation openness) and demographic factors (age, gender, technology generation) impact the risk perceptions associated with the integration of those technologies [27, 29].

In the last few decades, a rich knowledge base about potential acceptance drivers and resistance patterns has developed in academic research in many technical contexts [e.g. 30, 27]. Characteristically, however, there is an enormous gap on how to practically use acceptance knowledge in the rollout process: technical planners as well as people in charge in the communities often do not know how to identify acceptance factors and if so, when to integrate the public in the planning process [31]. When it comes to the integration of the public in the technology acceptance discussion, practical planners typically face two very contradicting positions. The first is the citizens' claim for the integration into the planning process. The justification of this claim is the argument that the public is the one that is confronted with infrastructure decisions and technological changes in the end and should thus be involved in the decision-making. The second position is typically used by experts (technical, political, legal) arguing that laypeople cannot make reasonable and reliable decisions due to their limited domain knowledge [32]. Also, it is a widespread prejudice that open discussions about a novel technology naturally lead to protest or disapproval – at least in early stages of the developmental process of the technology [26].

A further misunderstanding is that acceptance more or less refers to the mere convincing of the public at the very end of the planning and implementation process. This way, acceptance is used as a kind of top-down legitimation, a mere information act, without really asking the public about their perspectives. This engineering or marketing interpretation of acceptance is, however, not necessarily successful, as can be seen by the increasing number of public protests in the energy and mobility context, but also by the decreasing trust in the credibility and reliability of decision makers and public authorities [27, 33, 34].

From a social science point of view, acceptance should be treated, understood, and used differently. The approach should focus on the understanding of the perceptions of benefits and barriers of novel technologies in order to develop individually tailored information or communication strategies and integrate the user in technology development [32]. In this context, the public can take on different roles. According to [35], "public communication," "public consultation," and "public participation" are to be distinguished. In all forms, the public is informed by the technical planners. However, only the participation form allows all stakeholders and persons involved to be an active part in the planning process. Thus, to make use of public acceptance knowledge, it is essential to

1. Identify stages in the development cycle where integration of social acceptance is useful and in which form (all stages, different forms according to stage).
2. Empirically collect stakeholder perspectives on the new technology. Characteristically, these perspectives might differ from each other, might stress different pro- as well as contra motives and might be related to different economic, legal, political or societal issues. All are relevant. Only if the full picture is discussed and communicated within and across stakeholder groups, the mental and real transformation process can be successfully completed.
3. Feed the understanding and acceptance collected from the public as a valuable information source back into the technical development process (using any degrees of freedom) and, by this, creating a deeper understanding of the different sides.

For this, a holistic and integrative empirical methodology is required that bridges the gap between research and a practically oriented transfer of acceptance research into early stages of the development of a technology. Typically, a combination of qualitative and quantitative social science procedures is essential. While at the beginning qualitative procedures are useful (interview studies, focus groups and public hearings), the later stages of the implementation process require quantitative procedures and include hands-on experience (field studies, experiments).

In the following, we report on a citizens' dialogue that has been undertaken in the project APEROL[1] [36, this book] as an example of user requirements

---

[1] APEROL – Autonomous, Personalized Organization of Road Traffic and Digital Logistics", funded by the German Federal Ministry of Transport and Digital Infrastructure (BMVI, funding code 16AVF2134B)

empirically collected and later integrated. In the respective project, the task of the social science is to capture the acceptance-relevant requirements for automated traffic from the citizens' point of view, to identify user profiles and to collect context-adaptive usage conditions and new mobility services desired by the citizens. Together with the City of Aachen, Germany, as the communal project partner, the authors group at the Human-Computer Interaction Center (HCIC) at RWTH Aachen, Germany, is responsible for public communication, technology acceptance research, and the involvement of citizens in the project.

1.3 Citizen Dialogue - the collaboration of academia and the public

Increasingly, city participation approaches are used for municipal innovation management in the context of energy, infrastructure and mobility [37, 38, 39]. In a green paper, published by the European Commission [40], Citizen Science is referred to as a novel research agenda in Europe and describes a systematic and close connection of science and the general public. The idea of the link between science and the public is that both sides – the citizens and the scientists – acquire new knowledge, novel perspectives and a deeper understanding of complex topics which are addressed scientifically. This way, democratic interactions between and mutual understanding of science, society and public policy are improved. Also, the overall public knowledge is increased which, in turn, might empower laypeople in making informed decisions regarding socio-technical innovations.

In the course of the project cooperation, HCIC and the City of Aachen conducted a municipal citizens' dialogue with the support of all project partners. The overall target was to collect different stakeholder perspectives on autonomous driving and to shed light on potentially different perspectives with respect to the perceived benefits and drawbacks. In order to involve future users of all kinds, a citizens' science approach was used.

Overall, about 70–100 citizens were invited to participate in a day-long research session[2] about autonomous driving, future mobility strategies and their wishes and requirements for an autonomous shuttle service. All participants responded to an official announcement that invited interested citizens for an interactive workshop day with the topic of autonomous driving. All persons volunteered to take part and were gratified for their efforts with a small present. In order to demonstrate the joint efforts and the open communication policy, responsible mobility planners and communal policy makers of the city of Aachen as well as the interdisciplinary APEROL research team actively participated in the citizens hearing.

The day was structured in six thematic sessions (see Fig. 1.1) for a full understanding of all perspectives and to layout lines of argumentation from the citizens. Hereby, a mixed-method approach was used: qualitative methods like

---

[2] April 2019, Aachen, Germany.

focus group discussions and quantitative methods such as questionnaires as well as case studies were mixed in order to have a broad evaluation on the one side, but also to dive deep into specific topics on the other side. The participants were split into groups of six to seven people. Each group was accompanied by an expert moderator and external experts to the topic. The groups were changed for each new thematic session.

Photo 1 and 2: Citizens' dialogue, April 2019 in Aachen.
(©Human-Computer Interaction Center, RWTH Aachen)

The dialogue provided the opportunity to help shape current and future developments of the APEROL project and movements in the context of mobility in the City of Aachen.

1.4 Mixed Method Approach: Understanding Lines of Argumentation

As described, the research day was structured in six thematic sessions (see Fig. 1.1). The sessions included a mix of qualitative (e.g. task-based focus groups) and quantitative (e.g. questionnaires) methods to collect all perspectives on the topic at hand.

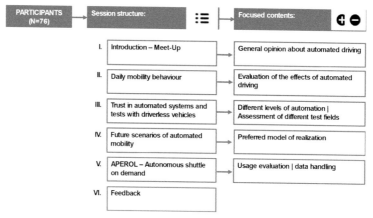

Figure 1.1: Overview of session structure and focused contents
of the citizens' dialogue (N=76).

The first session was defined as "Meet-up", the structure of the day was introduced and the general opinion of the participants about automated and autonomous driving was evaluated. Here, the use of autonomous driving for oneself, for society and the personal assessment of the technology were addressed. The second session was centered around the daily mobility behavior of the participants, questioning what impact autonomous driving might have for, e.g. possibilities of private car use, driving durations in the city or costs for transportation. The third session focused on the apostrophe participants trust in automated systems and also their personal assessment of test fields (e.g. in the inner city vs. on defined test environments). Following this, the fourth session looked into future scenarios of autonomous with different case studies and identified the participants preferred level of realization. Also, questions of trust in and evaluation of the perceived competence of stakeholders regarding autonomous driving were questioned. A focus on on-demand shuttle services was given in session five. Here, requirements, challenges and doubts concerning data handling were discussed.

1.5 Questions Addressed and Logic of Empirical Procedure

Based on the state of research and project-based goals, this article addresses:

1. **Attitudes:** How do citizens evaluate autonomous mobility in general, and, more specifically, for their own city?
2. **User specifics of on-demand mobility:** Which specific requirements do citizens have for on-demand autonomous shuttle service?

1.5.1 Sample, Data Collection and Analysis

Overall, N=76 participants from Aachen took part in the citizens' dialogue. In total 53.2% of the participants were male (n=41), 45.5% were female (n=35) and 1.3% did not decide. The average age was 47.1 years (SD=19.4) and varied between fourteen and 82 years. Therefore, the sample was older than the average of the German society with 44.3 years on average (Bundesamt 2017) [44]. The sample showed a variety of educational attainment and employment (see Table 1.1).

Table 1.1: Educational levels of the sample (N=76)

Education	Mean/Quota
No school diploma	3.9%
Secondary school certificate or less	5.2%
Vocational training	7.9%
High school diploma (A-Levels)	17.1%
University degree or higher	61.9%

In the following, we report on descriptive statistics (reporting means (M) and standard deviations (SD)) for the questions that were answered by the participants.

1.5.2 Mobility Behavior and Attitudes towards Autonomous Driving

In their general opinion about automated driving, 37% stated that they feel positive about autonomous driving, 7% even very positive. A further 37% stated a neutral feeling, 18% answered that they feel critical and 1% very critical about autonomous mobility.

To gain a deeper understanding, the participants were questioned if the impact of autonomous driving will be positive (or negative) for themselves, for society in general and if they feel optimistic (or pessimistic) about the introduction of the technology in everyday life.

The answers also show a positive attitude towards the future technology with an evaluation of M=3.47 (SD=1.06/5 = max. positive influence) for themselves and M=3.53 (SD=1.04) for society. Out of 65 participants, 47.7% feel (very) optimistic, 33.8% rather neutral and 18.4% (very) pessimistic about the technology.

Following these insights, a closer look at the mostly positive evaluation must be taken, to understand underlying evaluation concepts better. As can be seen from Fig. 1.2, the majority of the participants perceived autonomous shuttle services as meaningful (M = 3.83, SD = 0.86), useful (M = 3.79, SD = 0.88) and beneficial (M = 3.63, SD = 0.92). They also stated that they are in favor of the integration of the technology (M = 3.78, SD = 0.95) into daily mobility.

Questioning their own perceived comfort at different levels of automation, a diverse picture emerged (see Fig. 1.3). Here, the participants showed the tendency to feel less comfortable with higher levels of automation. Level 2 features were

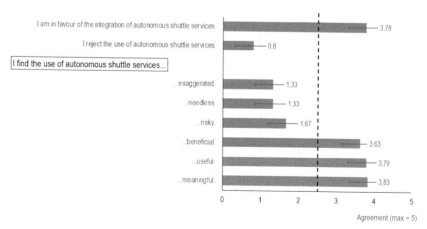

Figure 1.2: Evaluation (N = 76) of the use of autonomous shuttles (agreement scale max 5).

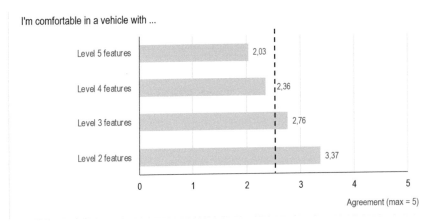

I'm comfortable in a vehicle with ...

Figure 1.3: Average evaluation (N = 76) of perceived comfort for different levels of automation (Agreement scale max 5).

overall rated with M = 3.37 (SD = 0.87), whereas Level 5 features scored M = 2.03 (SD = 1.27) regarding the feeling of comfort.

1.6 Requirements: On-demand Shuttle-Service

The participants were questioned about specific requirements for an autonomous on-demand shuttle service. The service was introduced via informational text and included the definition of autonomous driving according to the SAE automation definition [41] Here, the waiting time for a shuttle was an issue alongside the distance the participants would be willing to cover to get to a shuttle (pick-up point) and from the shuttle to the final destination (drop-off point).

The waiting time and distance were clustered in two categories, depending on the use-case: autonomous shuttle use for leisure (free time) or work and are results of the questionnaire handed out to each session. The task at hand was to give the most accurate time (in minutes) and distance (in meters) to be acceptable and unacceptable regarding the use of an autonomous shuttle-service.

The participants (n = 46)[3] were willing to wait 9.9 minutes (SD = 7.9) in their free time, but more than 20 minutes were perceived as too long to use the shuttle service for this purpose (M = 21.6, SD = 11.7). The tolerance of waiting shortens depending on the usage purpose. For work, the participants were willing to wait 6.9 minutes (SD = 4.3), but more than ten minutes were perceived as too long (M = 12.8, SD = 5.1) for this purpose.

Similar results can be displayed for the approved distance to cover: In their free time, an average distance of 529.6 (SD = 324.3) meters to the drop-off/pick-up point was too far away, while 287.66 meters were acceptable (SD = 229.1). The maximum distance to cover for a drop-off/ pick-up to use an autonomous

---

[3] Deviating number of participants due to not handed in scores.

shuttle for work purposes was 240.4 meters (SD = 165.7), whereas 430.6 (SD = 278.9) was perceived as too far away to use it. The tolerance of distance of drop-off/pick-up points for work purposes displays the actual distance of public transport stations in the city of Aachen.

1.7 Technology Usage, Test Fields and System Specifics

Considering the research focus of specific requirements for an on-demand autonomous shuttle service, especially, sessions three and four provided the possibility to gain insights into usage reasons, possible and acceptable research locations and system specific data security perceptions.

1.7.1 Usage Reasons

First, the given usage reasons were discussed, collected and further evaluated (see Fig. 1.4). Here, the possibility to offer mobility options for mobility-impaired people was among the most important reasons to use an autonomous on-demand shuttle service (M = 4.24, SD = 1.03), followed by using it in one's own free time, for example to get to an event (M = 3.82, SD = 1.05) and connect the city to the country and back (M = 3.71, SD = 1.24).

The least favorable reason was to pick-up children from or to kindergarten or school (M = 2.93, SD = 1.59). Overall, all the reasons for usage were agreed to, which also means that the versatility of the application possibilities of an autonomous shuttle-service was not only seen but also embraced.

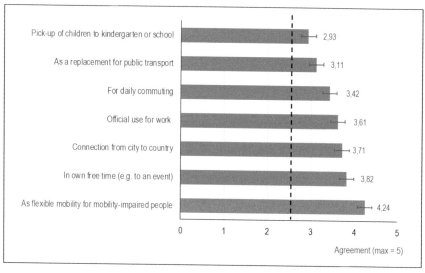

Figure 1.4: Average agreement (N = 76) to different reasons to use an autonomous on-demand shuttle (Agreement scale max 5).

1.7.2 Test Fields

It was also highly intriguing to see whether the interactive tests, which are necessary for the integration of the technology as a mobility option, are also viewed positively by the citizens.

Therefore, it was asked whether there are preferred test environments such as separate test tracks or even tests directly in the city. Here, too, all possibilities of different test environments were agreed upon: The separate test environments received the highest approval (M = 3.65, SD = 0.65, agreement scale: max. 5), alongside shorter predefined test tracks (M = 3.61, SD = 0.84). Followed by sparsely populated areas (M = 3.25, SD = 1.03) and on German highways (M = 2.91, SD = 1.39). Tests in the city were also approved but scored lowest with an average of M = 2.85 (SD = 1.37). This is a further indicator that is necessary to involve citizens in the choice of different test environments, to avoid unacceptable solutions.

1.7.3 System Specifics

The questions for system specifics focused on the willingness to share data with passengers or technology. Figure 1.5 shows the evaluation of the participants' willingness to share different kinds of information with other passengers. Time of arrival (M = 3.41, SD=1.30) is perceived as the most accepted information to share, whereas the concrete pick-up location (M = 2.68, S = 1.51) received the lowest ratings, although it was still agreed upon to be acceptable.

When asked for the willingness to share data with the technology, a slightly different picture emerged (see Fig. 1.6).

Here, only the following information was agreed to be acceptable to share: namely, information on the connecting journey, e.g. connecting times for other

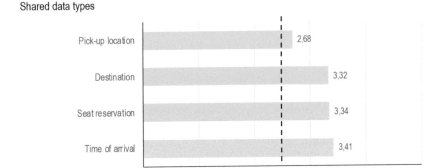

Figure 1.5: Average willingness (N = 76) to share various datatypes with other passengers (Agreement scale max 5).

Shared data types

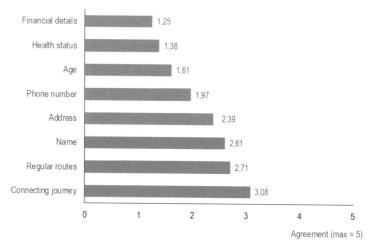

Figure 1.6: Average willingness (N=76) to share various types of data with the autonomous shuttle technology (Agreement scale max 5).

means of transport (M = 3.08, SD = 1.66), regular routes the passenger is taking (M = 2.71, SD = 1.50) and the passengers own name (M = 2.61, SD = 1.56). The transmission of all remaining types of data was on average rejected, e.g. the passenger's address (M = 2.39, SD = 1.62) or financial details to cover the booked trip (M = 1.25, SD = 1.51).

1.8 Generalizability and Transferability: Citizen Science, Conclusion and Lessons Learned

All the projects that are undertaken in many countries to promote autonomous driving are driven by the question of how far the respective results – may be limited to specific areas and exemplary test fields, cities, and countries – may be generalized and transferable to other cities and regions. Thus, it is important to foster solutions which will not only enhance mobility options in one selected city but illustrate with which software services and optimization strategies can be simulated with up-to-date maps to transfer them to other cities and regions.

With respect to the social science part – the findings regarding public perception and the methodology of citizens' participation – a high transferability is given for other cities and regions in Germany. Here, the national perspective on trust in planners and the general reluctance of data sharing processes in line with a high sensitivity of privacy issues is comparable across cities and geographic regions within Germany. However, beyond the applicability of results for German application areas, future studies and research efforts will have to focus on other nations, their mobility strategies and policies, which are not only very specific for the single European countries but also with respect to the cultural and societal practices of data storage and sharing in the automotive sector.

The presented research was directed at the integration of citizens in the future research on autonomous driving. Using a citizen science approach, the people of Aachen took part in a day-long discussion and study about possible ways to travel in cities in the near future. A focus on on-demand autonomous shuttle services was chosen to understand not only argumentation lines but further enable a deeper look at attitudes towards the technology as well as user specifics of on-demand mobility to help shape the technology development with the integration of those results into the development circle of the autonomous technology.

First and foremost, it can be pointed out that the citizens of Aachen were eager, interested and hands-on in their co-creational participation in the citizens' dialogue. A great openness towards new mobility solutions could be noted with which the citizens of Aachen are considering the role of shaping future mobility. They also showed high interest in participating in and shaping further studies, discussions and workshops.

As clearly voiced, citizens want to be part of a structured and transparent approach, dealing with data security and integration of technology in everyday life as well as in their mobility patterns. They further want the new mobility service to be an affordable, barrier-free technology. Also, "hands-on" experiences with automated and autonomous prototypes (urban living lab), participation in the development and the possibility of working on data protection concepts in urban space is a current wish of interested and well-informed citizens.

The citizens reported to have a positive and welcoming view of the technology, although, there are still obstacles that need to be dealt with. For example, privacy and data security was a perceived barrier in former research [15, 18, 14] and could be confirmed as a critical point in the citizens' dialogue. While potential passengers of an autonomous on-demand shuttle service were willing to share certain types of data with other passengers, e.g., their final destination or time of arrival, the willingness to share information with the technology was rather low. Consequently, data exchange should be viewed in a differentiated way and should not be generalized. The more personal the data was, the less willing the participants were to share them - again, validating earlier findings [15, 18].

However, it should be kept in mind that the reported findings relate to evaluations of (theoretical) data exchange scenarios. It will have to be validated in real driving scenarios in the future in how far the raised concerns about data exchange will hold especially when participants can really experience practical benefits and the comfort of autonomous driving.

Besides demands regarding data exchange and privacy, concrete requirements for an autonomous on-demand shuttle service could be revealed: The distance to a pick-up/drop-off point should not be further away than about 300 meters – mirroring the distance of the public transport stations in Aachen at the time of the study. Also, the accepted waiting time for the service could be narrowed down:

while in their own free time, the waiting duration might not be longer than twenty minutes to be approved, the tolerated time decreased down to seven minutes for a work-related trip.

Next to the specific results on requirements regarding data handling in the context of autonomous mobility and requirements regarding the autonomous shuttle itself, the citizens' dialogue as a method should also be critically discussed. The citizens' dialogue with the people of Aachen proofed to be a valuable source of information for the planners and scientists within the project. Overall, the public discourse about a novel technology in the context of mobility is not only a mandatory requirement of successful socio-technical transformations in which citizens can voice their perspectives during the technical development [26, 27]. Rather, public discourse and citizen participation are powerful instruments to form public knowledge and a guided understanding of the complexity of transformation processes. In addition, transparency and honesty in negotiating different positions is forming trust towards the planners, the communes and, last but not least, the transformation process [42, 27, 32].

Acknowledgements

The authors thank all citizens for their openness to share their ideas on a novel technology. In addition, this work was tremendously supported by the mission public and the city of Aachen. Specifically, we thank Kristine Hess-Akens and Uwe Müller for a joyful and vibrant cooperation between municipality and science.

This work has been funded by the Federal Ministry of Transport and Digital Infrastructure (BMVI) within the funding guideline, Automated and Connected Driving under the grant number 16AVF2134B.

Summary

Autonomous mobility transportation services represent innovative mobility solutions that enable cities and communes to provide efficient, sustainable, and safe mobility options. However, any technical innovation of this magnitude also generates initial frictions regarding public perceptions and the roll-out process. While, in general, autonomous driving is associated with a number of great advantages for the environment, safety and driving efficiency, providing many options for a timely and sustainable mobility, the loss of control over driving as well as data security and privacy issues are potential stumbling blocks for social acceptance among the population. For this reason, a central core element of any innovation management is to take users' perceptions into account early in the technology development and to develop sensitive public information and communication strategies.

This chapter describes a participatory citizens' forum in Aachen, Germany. Interested laypersons were asked about their perspectives on autonomous driving, social acceptance, privacy issues, data management, and rollout. The results show both, a great openness towards new mobility solutions, but also the claim of the public, to be part of the innovation process and the request for an open and transparent communication with respect to data security and the planned integration of novel technology into everyday life.

References

1. Breuer, J. J., A. Faulhaber, P. Frank and S. Gleissne. 2007. Real world safety benefits of brake assistance systems. In 20th International Technical Conference on the Enhanced Safety of Vehicles (ESV) (No. 07-0103).
2. Domin, C. 2010. Enabling safety and mobility through connectivity. *SAE International Journal of Passenger Cars – Electronic and Electrical Systems*, 3(2010-01-2318), 90–98.
3. Reimer, B. 2014. Driver assistance systems and the transition to automated vehicles: A path to increase older adult safety and mobility. *Public Policy & Aging Report*, 24(1), 27–31.
4. Yang, J. and J.F. Coughlin. 2014. In-vehicle technology for self-driving cars: Advantages and challenges for aging drivers. *International Journal of Automotive Technology*, 15(2), 333–340.
5. Calero Valdez, A. and M. Ziefle. 2019. The users' perspective on privacy trade-offs in health recommender systems. *International Journal of Human-Computer Studies*, 121, 108–121.
6. Schomakers, E.-M., C. Lidynia and M. Ziefle. 2019. Internet users' perceptions of information sensitivity - Insights from Germany. *International Journal of Information Management*, 46, 142–250.
7. Woolf, N. 2016. Tesla fatal autopilot crash: Family may have grounds to sue, legal experts say. Retrieved at March 2nd, 2020 from: The Guardian. URL: https://www.theguardian.com/technology/2016/jul/06/tesla-autopilot-crash-joshua-brown-family-potential-lawsuit.
8. NTV. 2019. "Autopilot" verhinderte Tesla-Crash nicht. Last visited: 05.07.2019. URL: https:IIWWN.n-tv.deIWirtschaft/Autopilot-verhinderte-Tesla-Crash-nicht-article 21031814.html.
9. Nguyen, D.H., A. Kobsa and G.R. Hayes. 2008. An empirical investigation of concerns of everyday tracking and recording technologies. Proceedings of the 10th international Conference on Ubiquitous Computing, pp. 182-191. ACM.
10. Morsing, M. and M. Schultz. 2006. Corporate social responsibility communication: stakeholder information, response and involvement strategies. *Business Ethics: A European Review*, 15(4), 323–338.
11. Renn, O. 1998. Three decades of risk research: Accomplishments and new challenges. *Journal of Risk Research*, 1, 49–71.
12. Hess, D.J. 2018. Social movements and energy democracy: Types and processes of mobilization. *Frontiers in Energy Research*, 6, 135.
13. Banks, V.A. and N.A. Stanton. 2016. Keep the driver in control: Automating automobiles of the future. *Applied Ergonomics*, 53, 389–395.

14. Schmidt, T., R. Philipsen and M. Ziefle. 2016. Dos and don'ts of data sharing in V2X-technology. *In*: Smart Cities, Green Technologies, and Intelligent Transport Systems. pp. 257–274. Springer, Cham.

15. Brell, T., R. Philipsen and M. Ziefle. 2019a. sCARy! Risk perceptions in autonomous driving – The influence of experience on perceived benefits and barriers. *Risk Analysis*, 39(2), 342–357.

16. European Commission. 2018a. Annual Accident Report 2018. Last visited: 04.07.2019. URL: https://ec.europa.eu/transport/road_safety/sites/roadsafety/files/ pdf/statistics/ dacota/asr2018.pdf.

17. Rovira, E., A.C. McLaughlin, R. Pak and L. High. 2019. Looking for age differences in self-driving vehicles: Examining the effects of automation reliability, driving risk, and physical impairment on trust. *Frontiers in Psychology*, 10.

18. Brell, T., H. Biermann, R. Philipsen and M. Ziefle. 2019b. Trust in autonomous technologies: A contextual comparison of influencing user factors. *In*: A. Moallem (Ed.). HCI for Cybersecurity, Privacy and Trust, HCII 2019, LNCS 11594, pp. 371–384. Springer Nature. Switzerland.

19. Raven, R.P., R.M. Mourik, C.F.J. Feenstra and E. Heiskanen. 2009. Modulating societal acceptance in new energy projects: Towards a toolkit methodology for project managers. *Energy*, 34(5), 564–574.

20. Sovacool, B.K., J. Kester, L. Noel and G.Z. de Rubens. 2018. The demographics of decarbonizing transport: The influence of gender, education, occupation, age, and household size on electric mobility preferences in the Nordic region. *Global Environmental Change*, 2, 86–100.

21. Canitez, F. 2019. Pathways to sustainable urban mobility in developing megacities: A socio-technical transition perspective. *Technological Forecasting and Social Change*, 141, 319–329.

22. Sheller, M. 2004. Automotive emotions: Feeling the car. *In*: Theory, Culture & Society 21.5, 221–242.

23. Fraedrich, E. and B. Lenz. 2014. Automated driving: Individual and societal aspects. *Transportation Research Record*, 2416(1), 64–72.

24. Josten, J., T. Schmidt, R. Philipsen, L. Eckstein and M. Ziefle. 2017. Privacy and initial information in automated driving—Evaluation of information demands and data sharing concerns. *In*: 2017 IEEE Intelligent Vehicles Symposium (IV), 541–546.

25. Wüstenhagen, R., M. Wolsink and M.J. Bürer. 2007. Social acceptance of renewable energy innovation: An introduction to the concept. *Energy Policy*, 35(5), 2683–2691.

26. Zaunbrecher, B. and M. Ziefle. 2016. Integrating acceptance-relevant factors into wind power planning. A discussion. *Sustainable Cities and Society*, 27, 307–314.

27. Arning, K. and M. Ziefle. 2020. Defenders of diesel: Anti-decarbonisation efforts and the pro-diesel protest movement in Germany. *Energy Research and Social Science*, 63, paper no. 101410, https://doi.org/10.1016/j.erss.2019.101410.

28. Gupta, N., A.R. Fischer and L.J. Frewer. 2012. Socio-psychological determinants of public acceptance of technologies: A review. *Public Understanding of Science*, 21(7), 782–795.

29. Ziefle, M., S. Beul-Leusmann, K. Kasugai and M. Schwalm. 2014. Public perception and acceptance of electric vehicles: Exploring users' perceived benefits and drawbacks. *In*: A. Marcus (Ed.). DUXU 2014, PART III: Design, User Experience, and Usability. LNCS 8519, pp. 628–639, Springer International Publishing Switzerland.

30. Assefa, G. and B. Frostell. 2007. Social sustainability and social acceptance in technology assessment: A case study of energy technologies. *Technology in Society*, 29(1), 63–78.

31. Stadelmann-Steffen, I. 2019. Bad news is bad news: Information effects and citizens' socio-political acceptance of new technologies of electricity transmission. *Land Use Policy*, 81, 531–545.

32. Schwarz, L. 2020. Empowered but powerless? Reassessing the citizens' power dynamics of the German energy transition. *Energy Research & Social Science*, 63, 101405.

33. Linzenich, A., K. Arning and M. Ziefle. 2019. Identifying the "Do's" and "Don'ts" for a Trust-Building CCU Product Label. *In*: Proceedings of the 8th International Conference on Smart Cities and Green ICT Systems. Volume 1: SMARTGREENS, pp. 58–69, Scitepress.

34. Leonidou, C.N. and D. Skarmeas. 2017. Gray shades of green: Causes and consequences of green skepticism. *Journal of Business Ethics*, 144(2), 401–415.

35. Rowe, G. and L.J. Frewer. 2005. A typology of public engagement mechanisms. *Science, Technology & Human Values*, 30, 251–290.

36. Prestifilippo, G. (forthcoming, this book). Smart services for autonomous mobility through logistics. *In*: G. Dartmann et al. (Eds., forthcoming). Intelligent Transportation: Mobility Through Artificial Intelligence for Autonomous Driving. CRC Press.

37. Michels, A. and L. De Graaf. 2010. Examining citizen participation: Local participatory policy making and democracy. *Local Government Studies*, 36(4), 477–491.

38. Carpini, M.X.D. 2020. Public Sector Communication and Democracy. *The Handbook of Public Sector Communication*, 31.

39. Peters, M.A. 2020. Citizen science and ecological democracy in the global science regime: The need for openness and participation. *Educational Philosophy and Theory*, 52(3), 221–226.

40. Scocientize. 2013. Green Paper on Citizen Science: Citizen Science for Europe – Towards a better society of empowered citizens and enhanced research. *In*: socientize. eu. Socientize, 2013, retrieved March, 2nd 2020.

41. SAE. On-Road Automated Vehicle Standards Committee. 2014. Taxonomy and definitions for terms related to on-road motor vehicle automated driving systems (SAE Standard J3016 201401). Warrendale, PA: SAE International.

42. Linzenich, A. and M. Ziefle. 2018. Uncovering the Impact of Trust and Perceived Fairness on the Acceptance of Wind Power Plants and Electricity Pylons. Proceedings of the 7th International Conference on Smart Cities and Green ICT Systems, Volume 1: SMARTGREENS, pp. 190-198, Scitepress.

43. Bundesamt (2017). Demopgraphics in the German polulation. https://www.destatis.de/DE/Themen/Querschnitt/Demografischer-Wandel/_inhalt.html, retrieved march 2021.

Artificial Intelligence for Autonomous Vehicle Fleets: Desired Requirements, Solutions and a Best Practice Project

Giovanni Prestifilippo

Managing Director PSI Logistics GmbH, Berlin, Germany

2.1 Smart Services for Autonomous Mobility in Logistics

Introduction

With the technological upheavals and ecological requirements, a sustainable paradigm shift is currently taking place in the transport sector. The conventional structures of the transport world are in upheaval. With the change in drive concepts, this shakes the production and value chain along the entire automotive industry. On the other hand, the development of driverless transport systems poses completely new challenges for transport infrastructure and control. This fundamental situation requires the development of entirely new mobility concepts. Some necessary solutions originate from logistics. In the logistics environment, decisive hardware and software components have already been implemented in line with the requirements of digitization and the innovation projects Industry 4.0 and Logistics 4.0. On the hardware side, automated-guided vehicles (AGV) have long been part of the intralogistics and production supply when transporting goods and materials (). On the software side, fleet management and the intelligent, route-optimized and increasingly decentralized control of the AGV are based on different system approaches such as real-time location and swarm intelligence (). In areas such as transport optimization, intelligent sensor technology and process

Email: prestifilippo@gmx.de

automation, logistics has proven to be an ideal test field, especially for the use of algorithms based on artificial intelligence (AI).

Current study and investigation results underline the necessity to develop new mobility concepts for public spaces from different points of view. A study presented by BVL.digital and HERE Technologies at the German Logistics Congress in Berlin in October 2019 [1], for example, highlights the improvement of traffic flow as an important element of urban environmental policy. For the study, 58 billion traffic data points were analyzed and 400 logistics experts from the network of the Federal Logistics Association were interviewed to measure city traffic. Result: On the main transport arteries of the major German cities, the speed is barely faster than 35 km/h – in the afternoon rush hours in Berlin and Frankfurt, Germany, for example, the speed is only 17 km/h. The consequences: increased pollutant emissions. As a direct effect, according to the study, a higher speed in cities significantly reduces the emission of carbon dioxide, nitrogen oxide and particulate matter. As an indirect effect, a higher speed may allow us to save vehicle capacities in CEP traffic. In a model calculation for Stuttgart, for example, an average speed increased by 5.7 km/h shows a reduction in emissions of up to 14 percent. The analysts see one of the biggest problems in a deficient traffic management and control system. In addition, an intelligent, digital traffic light setting is required [1].

Another source of assessment is the annual ranking "Global Competitiveness Index" of the World Economic Forum (WEF) in Davos [2]. In its current version, Germany continues to occupy the top position in terms of innovative capacity, the number of registered patents or scientific publications and their implementation into products. In terms of international competitiveness, however, the Federal Republic of Germany has dropped from third to seventh place. When asked how well it adapts to digital business models, it reaches the ninth place. In the category "Adaptation to new information technologies (ICT)", the country only reaches 38th place.

Finally, a look at the implementation periods of projects in public spaces might help to illustrate this connection. Even "normal planning procedures" require many years of implementation. Projects in public spaces, be it the development of an industrial park, a logistics property or the transport infrastructure, meet the interests of empowered citizens with the participation of the general public and then the affected public. Here, it is important to provide information and clarification about the technological, economic, organizational or even social consequences of the mobility transition at an early stage and to establish a comprehensive mediation process in advance of the project to avoid longer-term, unprofessional blockades.

Against this background, the development and piloting of innovative technology applications and concepts plays a particularly exposed role in the economic as well as in the ecological and social context. With the research project "APEROL – Autonomous, Personalized Organization of Road Traffic and Digital Logistics", the Federal Ministry of Transport and Digital Infrastructure (BMVI) has launched support for a joint project that focuses on the concept, development and validation of flexible, autonomous transports of people and goods in urban areas based on an "on-demand" mobility concept. With its development aspects and scenarios, the research project aims at the corresponding development of vehicles, software and mobility services as well as their evaluation in a pilot project first on a test facility and subsequent validation in public spaces of an urban transport infrastructure. This is accompanied by the optimized interpretation of suitable participation and communication concepts for wide adoption and public acceptance of the technology in society.

During the implementation of individual work packages, the project partners in the APEROL research project contribute basic components of their product portfolio and reflect corresponding experiences and principles from logistics.

2.2 The Research Project APEROL

In October 2018, PSI Logistics GmbH, Berlin, Germany, received the allocation decision from the Federal Ministry of Transport and Digital Infrastructure as consortium leader for the research project "Autonomous, Personal-Related Organization of Road Traffic and Digital Logistics" (APEROL) [3]. The research project will develop vehicle technology and software for autonomous driving in public spaces by the end of 2020. The research project is dedicated to the holistic approach based on new cloud and communication technologies for an optimized fully autonomous transport, which considers the individual needs of citizens and represents services for a fully autonomous transport system optimized to the needs of citizens. Under the project management of PSI Logistics GmbH, five research institutes of RWTH Aachen University and Trier University of Applied Sciences, the City of Aachen, Germany, and e.GO Mobile AG, MAT.TRAFFIC GmbH and Ergosign GmbH are developing the legal and technological basis for autonomous passenger transports and digitalized services in inner-city logistics.

PSI Logistics is responsible for the corresponding software area in the distribution of tasks among the project partners. To this end, the software company will develop suitable algorithms and simulation techniques for mobility services. This includes the creation of a user-friendly interface for intelligent mobility as well as the development of a transport control system and a comprehensive booking and remuneration system in a cloud-based framework.

At the end of the year 2019, the first pilot project should be executed on the test facility of the project partners. e.GO AG develops the model vehicles for autonomous driving. The city of Aachen, Germany, is responsible for the conceptual planning of the inner-city infrastructure.

The Transport Management System PSItms from the PSI Logistics Suite serves as the basis for the provided order planning system which will be used to plan, route and bill transport orders for autonomous cars. In the context of the research project, methods and procedures of AI will be developed and integrated into the scope of services of PSItms.

The special feature of the research project is its integrated approach. The conditions and prerequisites for the use of autonomous shuttle buses in public spaces are currently being researched in numerous projects worldwide. As a rule, vehicles for two to 15 persons with speeds between 20 and 50 km/h form the basis. In all projects, the use of the vehicles is designed for special scenarios. At the same time, leading OEMs such as Tesla, Google and Uber equip their vehicles with advanced assistance systems for autonomous driving – however, with limited use of driving functions.

The APEROL research project, on the other hand, develops autonomous driving without time limits. Traffic optimization and management is currently based almost exclusively on collective traffic control (e.g. traffic lights and change direction signs), i.e. on the adjustment of traffic flow capacities and (very limited) route patterns. Traffic information from the public media also has an influence. With new, optimally organized mobility services, which are entirely or partially based on autonomous vehicle fleets, knowledge about the current state of dynamic traffic in the road network will be much more comprehensive and precise – and the integrated vehicles behave exactly as calculated by a central fleet management system. The replacement of driver-determined traffic scenarios by concepts of autonomous vehicle fleets opens up completely new possibilities for traffic management – and not least promotes the traffic flow.

An important perspective of the project is the sustainable conceptualization and the transferability to other cities and regions. To achieve this goal, the consortium is developing planning software that can simulate services and optimization strategies with up-to-date maps and traffic data and thus transfer them to other cities and regions.

In view of the fundamental character of APEROL's research results, an application from the same partners to further develop the research project for autonomous trucks has now been approved.

2.3 The Partners and Their Expertise

The consortium covers all necessary competences.

- **e.GO Mobile AG** develops and produces innovative electric vehicles for cities and conurbations in Aachen, Germany. It was founded in 2015 as the successor to Streetscooter (acquired by Deutsche Post) from the RWTH Aachen University and combines leading university know-how with industrial practice. The e.GO Mobile is currently developing various new vehicle models to expand its product range – from the electromobilee.GO Kart to lightweight M1 class vehicles and autonomous PeopleMovers. These electric vehicles already offer automated driving functions as the so-called automated basis. The basic solution has a complete sensor setup for automated driving. In addition, steering and drive system can be controlled and regulated via an interface. In addition to the hardware, there is also software available that maps the basic functions of automated driving. This includes localization, trajectory planning and tracking as well as object recognition for security concepts. With this portfolio, e.GO takes over the development and provision of the hardware platform for the transport of passengers and goods as well as the organization of the multi-stage pilot operation in the APEROL project – and, with software interfaces to the sensor and actuator concept or the cloud framework, the integrated development of the automated driving functions in the hardware concept.

- The **City of Aachen**, Germany, has been involved in the project as an active partner from the very beginning and will enable real operation within the city. The vehicles of the e.GO partner offer the unique opportunity to evaluate the new services with "own" vehicles directly in a city. The department "Urban Development and Traffic Facilities" responsible for APEROL focuses on the challenge of autonomous/automated traffic and logistics in the project.

- The extensions of individual autonomous driving functions will be introduced into the project by the **PEM Institute** in close cooperation with e.GO. The chair "Production Engineering of E-Mobility Components" (PEM) at RWTH Aachen University is a new chair in the field of electromobility research.

- **PSI Logistics GmbH** develops standard software systems for logistics, especially in transport, warehousing, supply chain planning and the airport segment which process complex and comprehensive data in real time. Both logistics service providers such as Deutsche Post, Schweizerische Post and producers such as Bosch, Daimler, Würth and Tchibo use PSI solutions for their daily and strategic logistics tasks. In the research project, PSI Logistics especially brings in their knowledge and competence in the implementation of transport management systems for the control and planning of transports and resources which can be optimized according to different and complex parameters. They can be implemented as cloud solutions. Furthermore, the company contributes experience from the field of artificial intelligence (AI) to the project.

- The planning and optimization of cloud-based services is ensured by MAT. TRAFFIC and PSI as well as ISS and i5. The partners have many years of experience in the field of optimization processes for technical systems. The Environment Campus at the Institute for Software Systems of the **University of Applied Sciences Trier (ISS)** has completed various preliminary work in the field of software development for user interfaces, IT infrastructure and platforms. Among other things, ISS researches how innovative IT solutions can contribute to sustainable development and what effects IT has on people, the environment, the economy and society.

- The **i5** division (Computer Science/Information Systems 5) at RWTH Aachen University deals with the formal analysis, prototype development and practical testing of meta-information systems – concrete topics include mobile applications and services for mobility. With this specialization, i5 develops, in close cooperation with Ergosign, HCIC, PSI, PEM and ISS, the query control for the vehicles in the area of the GUI for intelligent mobility as well as the information systems for the booking and payment system of an intelligent mobility in urban scenarios. Via the open mobility platform (Mobility Broker) for digitalized mobility of i5, which is already in use in the city of Aachen, Germany, and its transport companies, the department also offers a scientific interface whose integration into the project may prove helpful.

- The Chair of communication Science at the Human-Computer Interaction Center (**HCIC**) of RWTH Aachen University deals with interface usability and the technology acceptance, thus the citizens' willingness to accept and use innovations in the mobility sector. Also, the integration of users and the adaptive customization of public communication for innovations is focused. In the project, the HCIC takes over the empirical-experimental description of the acceptance-relevant requirements for automatic traffic from the citizens' point of view, identifies user profiles, collects context-adaptive usage conditions and new mobility services desired by the citizens. In close cooperation with Ergosign, HCIC created the interface design for the app. Together with the City of Aachen, Germany, it is responsible for public communication and the involvement of citizens. During the iterative evaluation of the pilot operation, the HCIC ensures the formative evaluation and reflects the results to the technical partners for the iterative optimization of the pilot operation. The services are implemented by Ergosign and PSI who have already implemented various apps for various services.
- ISS and ICE will contribute the necessary competence in the field of IoT and networking to the project. **ICE** has many years of experience in the field of wireless communication with a special focus on technologies of the physical layer and software development for corresponding embedded systems and is intensively researching the development of future communication networks. It will also act as an interface to the BMBF PARIS project which is developing novel platforms and machine learning methods for autonomous driving. The results can also be integrated into the APEROL project, creating synergies. **ISS** is a member of the Internet of Things Expert Group of the Digital Summit and has experience in the development of IoT platforms and cloud systems for the Internet of Things. Prof. Dartmann also has expertise in the development of algorithms for real-time systems and experience in the field of optimization.
- The Aachen-based technology company **MAT.TRAFFIC GmbH** is a service provider developing innovative concepts and architectures as well as methods for modelling and controlling traffic and software solutions with a focus on "cooperative systems". It has special expertise in local traffic light control (energy efficiency), modelling of traffic as well as simulation of traffic scenarios and network effects. Thus, MAT.TRAFFIC takes over the modelling and optimized control of autonomous (and other) traffic at a network-wide, strategic level in the research project.

- **Ergosign GmbH** is one of the pioneers for services in the areas of usability engineering and user experience design in the German-speaking market. In the APEROL project, the development services for human-machine interfaces which are adequate for cognition, tasks and context are applied.

In addition, the German Technical Inspection Association **TÜV** subcontracts the acceptance of the pilot operation to determine driving and traffic safety and the feasibility of automatic pilot operation. In addition, the course of the project includes obtaining **legal opinions** to assess the actual state (project start) and to approve the pilot operation developed in the project (upon project completion). **Campus AG, the German National Digital Summit** are also **represented** in the expert advisory board accompanying the project. **The latter** pursues the project as a test model for autonomous pilot operation with the interest to adapt essential developments.

2.4 Results-oriented Research with Scenarios

As consortium leader, PSI Logistics has especially brought the economic aspects of feasibility for the project results in the areas of vehicle development, software and infrastructure to the fore. To this end, scenarios were developed prior to the work packages, task areas were defined and assigned to project partners.

The usage scenarios were first used to consider the requirements of fully automated, inner-city passenger transport and goods transport. Results in the latter area certainly provide the basis for the above-mentioned research project on autonomous goods transport vehicles. In addition to the requirements of dynamic customer inquiries with advance booking intentions, individual and collective transport and time guarantees for time windows, the requirements emerged as questions of the vehicle type, the integration options of the traffic situation and various transport alternatives.

These also characterize the innovations in the research project because – as you can see in the example of the ridesharing service Moia in Hamburg, Germany [4] – app-based offers with specific information on the time window and near-position locations for largely individualized multi-person transports are already available on the market. In Hamburg, this even takes place with e-vehicles from VW production. The difference to the APEROL approach lies in the methodically innovative, holistic conception and optimization of the entire autonomous traffic including individual user needs, the current traffic situation, the residual energy of the electric vehicles and various environmental data. In its adaptation, the fleet management of autonomous vehicles is also quite suitable for covering the supply gaps of public transport in rural regions. This corresponds to the preferred reasons for using APEROL shuttles.

When defining specific requirements for the booking and remuneration system, it quickly became clear that fair prices had to be created on a multi-criteria basis within the planned framework. As the system offers more scope for optimization than conventional approaches, potential users can be offered a favorable price with high flexibility in time and place.

Regarding vehicle development, mobility services and infrastructure planning, APEROL provides a comprehensive system structure. In this structure, the e.GO Mover is first technologically tailored to the special requirements before being piloted for the first time at the test facility. In addition to a reduced engineering model, the e.GO Mover is equipped with a special camera sensor system for object classification and recognition of light signal systems. The equipment with laser scanners supports 3D environment detection for free space detection and positioning. The environment model is generated from freely available information (OpenStreetMap). Long- and short-range radar ensures object detection, DGPS ensures high-precision positioning.

For the test scenarios, three areas appropriate to the iterative development steps are used consecutively: the ATC Aldenhoven "factory site", the Campus Boulevard RWTH Aachen University and the designated routes in the Aachen city area. With the participatory and user-oriented approach across all phases of technology development, in addition, acceptance, communication and information needs are identified from the citizens' point of view and continuously integrated into the development process of the project. On 6 April 2019, for example, an international dialog with the citizens on autonomous mobility took place in Aachen, Germany, to record the demands of the population up to project piloting and to identify and react early to factors hampering acceptance. As a general attitude, the preferred reasons for using APEROL shuttles on a scale of 1 (not at all) to 5 (exclusively) are "flexible mobility for people with reduced mobility" (4.2), "transport in free time" (3.8) and the "connection between urban and rural areas" (3.7). In terms of acceptance of driverless vehicles in city traffic (2.74), however, there is an even greater need for information and clarification. The results of the APEROL research project will in the end provide a factual guarantee of this.

2.5 Innovation Potentials

Beyond to the overall view of the methodological innovation, which pursues an integral view with the development of highly automated driving, electric vehicles, the optimized alignment to individual user needs and an app with user-friendly GUI as well as the integration of the current traffic situation, the residual energy of electric vehicles and various environmental data, with which the results for autonomous vehicle fleets can be transferred to new scenarios (cities), the individual project partners each see specific innovation potential for their industry segment. For PSI Logistics, the focus is on the functional and technological expansion of the range of uses of the PSItms transport management system from the PSI Logistics Suite.

In the research project, PSI Logistics is pursuing the goal of occupying a completely new market for the organization and control of autonomous transports of persons and goods. With their solutions in transport logistics, PSI Logistics aims to be the market leader for autonomous driving in the context of electromobility. Furthermore, the aim of PSI Logistics together with the competence of the affiliated business unit PSI Energy in the PSI Group is to optimally design the loading and unloading of electric vehicles.

With today's PSI Logistics solutions, large logistics service providers such as Schweizerische Post or industrial producers such as Bosch design, control and optimize their transport networks. In the concerted interaction of modules and functionalities and the seamless, intelligent and conflict-free coupling of strategic supply chain network design and operative transport management of PSItms, Schweizerische Post covers the complete process sequence. The functional spectrum of IT ranges from optimum network design and pre-planning through the creation of reference routes, order acceptance, scheduling and resource planning, including the coordinated use of 6,000 trucks, time window management and cost minimization to transport execution, invoicing and processing and analysis of event data. With the warehouse management system PSIwms and PSItms, two IT systems from the PSI Logistics Suite also ensure efficient warehouse processes, optimum process control and demand-oriented production supply in e.GO's production.

In parallel to this, PSI Logistics is intensively pursuing the beneficial integration of Robotic Process Automation (RPA) as well as methods and procedures of Artificial Intelligence (AI) such as deep learning and neural networks into the standard works of the PSI Logistics Suite. With its development competence, which has won several awards in the past four years, the company is one of the innovation leaders in the software industry in Germany.

Within the scope of the work packages defined in the research project, PSI Logistics implements several requirements in a transport control system. This is provided as a cloud solution. It transfers the user data to the system via apps. In the transport control system, the recorded transport order data is dispatched to autonomous vehicles. Depending on the traffic situation in the area of application under consideration, the optimal transport routes including travel times are then determined.

The development takes place as a prototype on the existing and established product transport management system PSItms. With its contribution, the research project can already be built on many necessary basic functionalities. The PSItms program standard already includes algorithms for all necessary planning processes – from order entry to scheduling and resource management, pickup & delivery as well as optimized route planning considering restrictions such as specifications for driving times and rest periods. Using Tracking & Tracing (T&T) and dynamic route changes, an actual/target comparison of the planning processes can also be carried out in the TMS. Finally, PSItms is used for the billing procedure with prior cost calculation and subsequent allocation to points of payment which can support multi-criteria pricing. PSI Logistics thus offers a technology for the recording of transport orders, inquiries via smartphone, scheduling and individual route planning and billing that has already proven itself in the logistics environment.

The development of new software functionalities can thus be concentrated on the implementation of the core functions required for the project. One focus is on the cloud solution. Within the scope of the corresponding work package, new functionalities are implemented in PSI*tms*. Essentially, the focus will be on the prototype of the cloud framework and the interfaces to the practical partners as well as to the vehicle sensors. The core of the PSI*tms* program extension is to provide the functions for the complex parameters in the context of the task, to design and control them efficiently and optimally.

2.6 Outlook

The APEROL research project initiated by PSI Logistics is expected to provide important insights into the technical and economic implementation of traffic optimization and management with vehicle fleets of autonomous vehicles. The prototypical structure of a platform and the app that enable users to direct their transport requests (persons, goods, products) and to receive a solution highly valuable due to the implementation validated by the pilot operation. The prototype organizes transport port vehicles in a route-optimized way. Relevant experience for a practical autonomous vehicle mobility under logistic requirements with this complexity is invaluable. The technical and scientific evidence of the savings effects of networking will lead to an improvement in transport systems which will underline both ecological and economic sustainability. This will make it possible to demonstrate the economic benefits of autonomous mobility and, not least, to increase the acceptance of a broader market presence for transport systems with autonomous vehicles. The project results will also trigger significant improvements in the standardization of efficient mobility under logistic conditions.

With the development of newly acquired knowledge for the management and control of autonomous vehicles in both passenger and freight traffic, the prototype created in the research project will be incorporated into the PSI standard product PSItms (Transport Management System) in its short-term utilization. At the same time, the cloud solution will be marketed as a portal for recording transport orders. The medium-term exploitation of the solution resulting from the research project will be continued in the Aachen region – and its marketing in new regions in Germany will be further promoted in accordance with the project requirements of transferability. With the successful rollout of the solution components, PSI Logistics is also opening up a completely new, sustainable business area within the scope of digitalization with the goal of international market leadership in this area. In the following steps, the targeted identification of companies for autonomous fleet vehicles, such as automobile manufacturers, rental car companies and logistics service providers or completely new providers of transport services, is conceivable – first in German-speaking countries but gradually also far beyond. This means that the defined objectives of the consortium leader have been met.

References

1. Online: https://go.engage.here.com/Accelerating-Urban-Logistics.html
2. Online: https://www.weforum.org/reports/how-to-end-a-decade-of-lost-productivity-growth
3. Online: https://www.bmvi.de/SharedDocs/DE/Artikel/DG/AVF-projekte aperol.html
4. Online: https://www.moia.io/de-DE/hamburg

Large-scale Simulation of a Mobility-on-Demand System for the German City Aachen

Paul Mathias and Paul Dowideit

MAT.TRAFFIC GmbH, Heinrichsallee 40, 52062 Aachen, Germany

The German research project APEROL is dedicated to the implementation, testing and validation of a holistic approach for an optimized road traffic system that is organized as an autonomous mobility-on-demand (AMoD) service. As there are only very limited possibilities in real test sites, scientific investigation is also put on a realistic, large-scale simulation of the whole traffic of an exemplary German city (Aachen). The ultimate goal here is, starting from the traffic situation as it is in Aachen today, to simulate several possible future traffic scenarios and assess them.

The simulation tasks in APEROL are based on the modular open-source framework MATSim from the ETH Zurich. MATSim offers interfaces that allows specific extensions of the system; in the APEROL case to integrate autonomous shuttles (AS), which are controlled through a mobility-on-demand fleet management.

This chapter starts with a short overview about other big MATSim AMoD traffic simulations (of other cities) that comprises their motivation and goals as well as main results and limitations. In contrast to this, the special APEROL requirements are introduced, and the technical solutions are presented to address those requirements. What is outstanding here is the performance of the large-scale simulation for Aachen.

The simulation is shown step by step, on how to set up, configure and extend the MATSim simulation system. Extensions are mainly (1) the implementation of autonomous shuttles to model the AMoD subsystem, and (2) interfaces to

---

*Corresponding author: ??

couple the MATSim subsystem with a so-named Mobility Center that contains the AS fleet management. The AMoD related fleet management algorithms are not discussed in this chapter. A basic, less sophisticated, version of such an algorithm, which is designed for high computational performance, has been developed and integrated. The chapter concludes with first results of the AMoD simulation and a brief outlook to further simulation experiments.

3.1 Introduction

3.1.1 APEROL

APEROL stands for "Autonomous, personalized organization of road transport and digital logistics" and is a German research project, founded by the BMVI (Federal Ministry of Transport and Digital Infrastructure) [16].

The project is dedicated to the implementation, testing and validation of a holistic approach for an optimized autonomous transport with provision of suitable mobility services. The services are intended for individual passenger and freight transport, taking into account individual citizen needs and optimizations in terms of public acceptance. It is targeting two main goals that directly correspond with autonomous transport. The first one is related to the fundamental research of autonomous driving itself. For example, processing of sensor data, acknowledgement of the environment, interpretation of situations and the strategies of the autonomous driving maneuvers. The second goal comprises the development of a cloud solution including a complete and optimized autonomous traffic system that utilizes new communication technologies integrated with the management of the autonomous fleets (AMoD algorithms). These will provide APEROL an agent-based simulation of all traffic in the city of Aachen and to create a test bed for fleet management algorithms and optimization procedures. As a result, future traffic scenarios with different distributions of traffic modes can be evaluated for simulated and analyzed for feasibility, costs and transferability to other cities and regions can be determined.

Comprehensive and large-scale traffic simulations are at the core of this research. The general approach here is to model realistically the complete, individual and public, traffic of a medium size German city as it is today by means of an extended MATSim system, and then to investigate several possible future scenarios that use more or less extensively AMoD mobility services.

3.1.2 AMoD City Simulation with MATSim

MATSim [8], [10] is an open-source transport simulation framework that allows large-scale agent-based simulations, whereby agents represent the mobility related activities of real people in a road traffic system. For this, all agents possess a daily plan of intended activities which start at a specific time of the day, have defined durations, and are assigned to different transportation modes. The activities take place at different locations in the transportation network, and consequently, by following their activities, agents are moving through the network. The MATSim simulation framework comprises functional blocks which can be combined or used stand-alone. Most importantly, such modules can be extended or replaced by custom implementations to adapt the simulation to project specific requirements.

By default, MATSim allows the simulation of car traffic, public transit and slow modes, such as bicycling or walking. The outstanding challenge with the traffic simulation of those fleets is performance, especially with big traffic networks with numerous fleet shuttles. Road-based modes, such as private cars, are simulated in a time-step based manner in a network of queues with all participants at the same time. This way it is possible that bottlenecks and congestions emerge, and agents arrive late at their respective activity locations. In this research, not only the basic network simulation is utilized but also extended features like the replanning of agents plans to cover the agent's perception and behavior. One scenario in MATSim consists of multiple iterations, and each iteration consists of the traffic flow simulation and learning process as represented by the following formal cycle in Fig. 3.1. The number of iterations can be individually configured but should be repeated until the average population performance stabilizes.

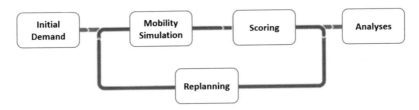

Figure 3.1: MATSim Cycle [8].

The travelers in the simulated transport network are modelled by a population of agents, each of them having specific travel plans assigned. The essence of such plans is the daily activity pattern of the agents, i.e. start and end locations, modes of transport and duration of stay at each location. Within each 24h simulation iteration, each agent tries to carry out its plan for the day. Various factors (traffic jam, no vehicles available, delays in public transport and etc.) could result in partial or complete failure to accomplish the plan. Since a scoring, combined with a replanning of the agent plans due to a co-evolutionary principle (mutation and recombination), is executed after each day iteration, agent plans tend to fit the supply of the transport system better from iteration to iteration. After several iterations there will be a convergence to a kind of equilibrium state, which then can be evaluated. The following table (Table 3.1) summarizes the mapping between simulation and reality.

In the context of this research, one transport mode shall be using autonomous shuttles on demand. These shuttles have very special requirements, as their routes and stops need to be controlled from an external mobility center. Also,

Table 3.1: Mapping between simulation and reality

MatSim		Reality	
Agent plans	• Contains for each agent his entire daily activities with locations, start/ arrival times, transport modes, etc.	Individual transport demand	• The overall traffic demand defined through activities of the road users.
Scoring of a simulation iteration	• Scoring by predefined evaluation functions (can be multi-objective).	User experience	• The user experience at this day.
Re-planning	• Based on the scoring depended re-planning (co-evolutionary principle).	Learning process of road users	• Behavior of road users gradually accommodates depending on their experiences.
Final iteration	• It is assumed that over the iterations the system converges to a state of best evaluation.	Optimal traffic flow	• Optimal scenario according to the possibilities and user preferences.

such shuttles are supposed to send status data periodically to the center. For all of this, the MATSim simulation framework has been extended by so-named AMoD shuttles, which can populate the traffic network as well. Each AMoD shuttle has an internal timetable for the near future. This timetable can be modified whenever a new travel demand appears that could be serviced by this shuttle. Furthermore, a shuttle could also have an idle time period which it spends in a waiting position at a fixed location (like a taxi) or by going to more suitable locations in the network with more demand (re-balancing). The sophisticated fleet management control algorithms (AMoD) are part of the APEROL mobility center. The AMoD part comprises the assignment of travelers / agents to shuttles and the optimized routes and stops of all AMoD shuttles (including re-balancing routes).

3.1.3 Related Research

The APEROL project investigates both the practical capabilities of autonomous vehicles and their application as a future means of transport, in large urban areas, in the context of a fully dynamic, demand-controlled bus operation (Autonomous transit on demand, ATOD). The assessment of the expected user acceptance has to take into account availability, time, reliability, accessibility and price, while offering the possibility to vary these parameters.

Autonomous vehicles can operate in different modes, ranging from pure point-to-point traffic with individual passengers [1], on the bundling of requests with similar start and destination regions (Shared Autonomous Vehicle, SAV) [2], up to ATOD models [3].

On the basis of recorded taxi rides in Manhattan, it was shown that dynamic carpools can reduce the total number of vehicle kilometers travelled [4], with more than one passenger on almost 100% of all journeys at peak times [5]. This research was based on the assumption that all such journeys are made as potential carpooling and that no substituted routes added.

Martinez and Viegas show that a combination of SAV and ATOD can meet the transport demand of a city like Lisbon [6]. However, they replace all private transport.

Bischoff and Maciejewski simulate the replacement of individual traffic, in the entire city area of Berlin, by autonomous taxis using the DVRP (Dynamic Vehicle Routing Problem) extension available in MATSim. This simulation was also without the choice of transport mode, by considering all current IV users as AV taxi users and removing all other users and transport modes from the model [7]. Under these conditions it seems possible to handle the journeys of 1.1 million IV users with 100,000 autonomous taxis.

The traffic simulation with MATSim [8] goes beyond the usual procedure [2, 3] where the procedure is mapped and implemented as an additional means of transport within MATSim. Instead, the core of the control and planning takes place in a mobility center connected via suitable interfaces. Thus, MATSim is transformed from a pure simulation environment to a test environment for traffic engineering procedures. Different mobility centers can be compared under constant laboratory conditions in different scenarios.

Simulations for larger areas, like the city of Aachen, require certain simplifications. For example, limiting the number of agents by random selection [9], reduced number of vehicles, appropriately scaled line capacities, etc. with subsequent extrapolation to the total [3].

Through the use of particularly powerful hardware and computationally intensive process steps (especially the optimization algorithms of the mobility center), it is possible to take into account every road user and every vehicle. This means that the individual choice of transport modes can also be taken into account, allowing conclusions about the design of bus operations organized as carpools, which is necessary for the greatest possible acceptance.

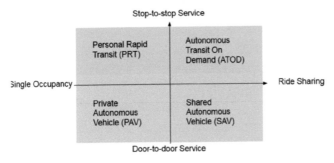

Figure 3.2: The four transport operation mode dimensions.

3.2 The APEROL System Environment

This sub-chapter describes how the overall simulation-based APEROL system is structured and technical approach. With regards to functionality, the overall system is sub-divided into two main blocks: (1) the APEROL Simulation (MATSim) and (2) the APEROL Mobility Center. Moreover, attention shall also be paid to the various MATSim extensions which were needed to address all requirements of the APEROL services (for instance AMoD shuttles).

3.2.1 System Overview

The overall APEROL system, which is based on a traffic simulation environment, is intended to represent current the real world (all relevant features and dynamics addressed) and to have the unaltered APEROL mobility related core services remain. The system consists of two subsystems (see also Fig. 3.3):

1. **APEROL Simulation**

    The simulation replaces the actual system with the transport system, the road users, and the different types of vehicles (transport modes). The simulation is realized by MATSim which is an activity-based, extendable, multi-agent simulation framework implemented in Java and available as open source [14].

2. **APEROL Mobility Center**

    The Mobility Center mainly comprises the booking process and the shuttle fleet management. The center is a simplified version of an actual Mobility Center that is used in the real test site of APEROL. Therefore, it must be considered as a container of the core service functionality relevant for the dynamics of the autonomous shuttle fleet within the traffic simulation part.

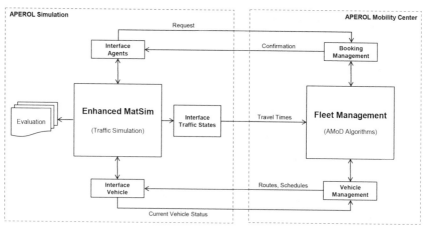

Figure 3.3: System overview of the APEROL simulation environment with two cooperating subsystems.

As far as AMoD shuttles are concerned, the two subsystems interact by periodically exchanging data. This is because the fleet management of the shuttles resides in the APEROL Mobility Center. The data exchange is handled by several interfaces which have been implemented. These interfaces are simplified versions of the APEROL interfaces in the real test site.

The simulation subsystem's operation is triggered by the Mobility Center. Both subsystems are synchronized through a 'step-forward' signal that is sent from the Fleet Management to the MATSim simulation. After receiving this signal, the simulation runs one simulation second further.

In the following subchapters the subsystems are presented in more detail.

3.2.2 The APEROL Simulation Subsystem

Until now, actual autonomous shuttle fleets of relevant size have not been available. Therefore, in order to investigate the dynamics and their integration into conventional traffic and performance related issues, a large-scale APEROL simulation subsystem has been used. This simulation part models the actual transport system, the road users and the different transport modes. It is realized by MATSim which is an activity-based, extendable, multi-agent simulation framework [14].

Extensions of the basic functionality of MATSim is important to the APEROL system because specific features of the autonomous shuttle operation need to be integrated. For this a control loop needs to be established that connects the MATSim agents with the Mobility Center (asking for a transport service), and the Mobility Center with the shuttles (to inform the center about the current position of the shuttles and to assign shuttles to agents / passengers and to determine shuttle routes). In addition, the simulation part shall send traffic state information (travel times per link) to the Mobility Center to support optimal, link state dependent routing.

3.2.2.1 Agent Interface

In the real test field, the inquiry of shuttle transport services is realized by a smartphone app utilizing dialogues between the client and the Mobility Center. The online booking dialogues can be quite sophisticated, since the client can receive, and have to choose between, multiple transport and pricing options. In the simulation subsystem a smaller interface is utilized based on the following simplifications:

- Each agent asks for service individually (no group inquiries). All such inquiries are based on the agent plans, which are derived from the traffic demand information.
- Agents inquire their transport ride randomly between 5 and 30 minutes before the desired pickup time.
- The agent receives only one offer from the Mobility Center.
- The agent does not need to confirm this offer. It is confirmed automatically.

3.2.2.2 *Autonomous Shuttles and their Interfaces*

The autonomous shuttle operation is controlled by the fleet management of the Mobility Center. The shuttle service comprises to shuttles with 7 and 15 seats. Autonomous shuttles are derived from standard MATSim vehicles with extended functionality that consists of the following:

- The internal database of each shuttle contains a transport schedule, which is primarily a sequence of stops (boarding and disembarking) and the routes between the stops. Those schedules can be modified remotely by the fleet management.
- The assigned agents/passengers enter or leave the shuttle at the stops.
- The interface to exchange data with the Fleet Management of the external Mobility Center.
- If the schedule is empty, then the shuttle remains at the last entered stop.
- The shuttle also drives without passengers (as long as its schedule is not empty).

Apart from this, the simulation environment handles shuttles the same as other vehicles (of different types); i.e. the shuttles are part of the overall traffic in the road network.

3.2.2.3 *Traffic States Interface*

The fleet management in the APEROL Mobility Center periodically calculates autonomous shuttle routes based on current network traffic states which are defined as travel time per network link. As a result, information needs to be determined dynamically in the simulation subsystem. The calculation of these travel times per link is already done by MATSim and only needs to be sent to the Mobility Center.

3.2.3   The APEROL Mobility Center

The APEROL Mobility Center comprises the infrastructure subsystems. However, for the simulation environment a scaled back version of the actual Mobility Center test site, designed for the needs of the simulation, is used. Only core functionality required to (1) Booking Management (transport requests of agents), (2) Fleet Management (the AMoD algorithms to assign requests to shuttles and adjust shuttle routes, and (3) Vehicle Management (the data exchange with the shuttles in the network. This Mobility Center structure is shown in the right part of Fig. 3.3. The dynamic exchange of data is realized by three interfaces that are part of the above described functional blocks:

Booking Management ↔ Agents

Fleet Management ↔ Simulation Network

Vehicle Management ↔ Shuttles

The simulation's Mobility Center is optimized to support large scale experiments, such as the simulation of traffic of an entire city traffic with reasonable performance.

3.2.3.1 Booking Management

The Booking Management block has the minimum functionality needed within the simulation environment. It receives all incoming service requests from the MATSim agents and forwards them to the Fleet Management block. All requests are one way (no confirmation) and accepted automatically.

3.2.3.2 Fleet Management

The Fleet Management, in the simulation context, is synonymous with the AMoD (AMoD = Autonomous Mobility on Demand) management for autonomous shuttles. Such AMoD services are designed to dynamically formulate the optimal strategy to serve all ride requests within a road network during the course of a day. Optimality, of course, depends on a given objective function, which might be multi-criterial. This is because there are at least two perspectives which might be, at least partially, in conflict with one another. The first is the clients' perspective, who might seek minimal wait and travel times. This is in a way the egoistic perspective. In contrast, there is the interest of the system operator, whose goal it is to maintain a reasonable business model and minimize operation costs (number of shuttles, kilometers driven, maintenance costs, etc.). Eventually, the competing goals might end up in a kind of equilibrium where the interests are balanced.

To cover all relevant aspects of the operation on a large scale, the corresponding optimization problems are very challenging. They are difficult to solve analytically, since, for instance, the shuttle-client assignment must be a regarded as integer linear programming [5], which is well known to be NP-complete.

In APEROL, the intention is to investigate the large-scale mobility scenario of Aachen, which is a medium sized city with more than one million ride requests per day. Performance requirements have resulted in the simplifications: (1) each shuttle offers its service individually, in response to client requests, with the fleet management selecting the best fit, (2) the shuttle-client assignment depends on the spatial proximity, and (3) no transfer procedures are allowed between shuttles.

Apart from those simplifications, the AMoD operation used in the APEROL simulation environment is characterized through:

- The shuttles in operation have two sizes in equal shares: 7 seats and 15 seats.
- The transportation requests are sent by the clients (agents) randomly between 5 and 30 minutes before pickup time.
- There is a defined trip time tolerance value rt ($0 < rt$) as a restriction for the shuttles to modify their existing routes. For every client centering a shuttle, an initial trip time tt_c^0 is determined on the basis of its route and the current link travel times. In order to pick up more clients, the shuttle may only modify its

current route if the following holds: $tt_c^0(1+rt) \geq t_c^{New}$, for all clients c in the shuttle.

- The client (agent) waits for the assigned shuttle and will not enter another shuttle.
- Group booking is possible. Bookings of groups of several passengers with the same origin and destination are possible.
- A conducted booking is fixed. A subsequent re-assigning of clients to another shuttle before pickup is not possible.

The AMoD procedure comprises three algorithmic parts: (1) shuttle-client assignment, (2) routing, and (3) rebalancing.

Shuttle-client Assignment

The shuttle-client assignment is the task to dynamically assign multiple requests to multiple shuttles of a certain capacity by taking into account the current schedules / routes of the shuttles. Each trip request consists of the time of request, a pickup time, pickup and drop-off location. The assignment strategy is distributed: Each shuttle (within a certain distance from the pickup locations) validates whether it is possible to incorporate one or more of the new requests, into its current schedule, without violating the trip time tolerances of the onboard passengers. If affirmative the shuttle sends its offer to a central dispatcher. But only those shuttles are asked to check their availability that have free capacity and will be at pickup time within a certain distance from the pickup locations. The dispatcher receives all shuttle offers and accepts those that fit into a given balance strategy of assignments.

The assignment algorithm on the dispatcher level – conducted every simulation second – is detailed as follows:

1. Collect all new trip requests from clients.
2. Query a selected set of shuttles to check their availability to serve the requests.
3. Collect the shuttle offers.
4. Determine the offers to accept in accordance with the given balance strategy.
5. Instruct the concerned shuttles to update their schedules accordingly.

On the shuttle level the algorithm can be described like:

1. Receive the new trip requests from clients that are within a certain distance of the shuttles current position.
2. Check whether there is a subset of requests that can be served by determining all possible detours of the current route that are in line with the trip time frame tolerances of all passengers.
3. Formalize all identified request subsets that can be served together with their route modification.
4. Send the results to the dispatcher.

Finally, it is pointed out that the whole shuttle-client assignment computation is a parallel scheme. It is to a far degree scalable, as it can run on any number of processor cores. The PC hardware of the APEROL simulation environment consists of 64 cores.

Routing

The shuttle routing is used extensively within the AMoD algorithm to do the optimal shuttle-client assignment and to update the current shuttle schemes. Therefore, it is based on precalculated optimal route-trees for the whole network, which are based on the current link travel times. Two different types of trees are calculated: (1) all the routes that go from any network node to a fixed node, and (2) all the routes that go from a fixed node to any other network node. If link travel times change significantly, as a consequence of changed traffic states, the corresponding route-trees are updated. The approach of looking up the routes in a precalculated database drastically improves the performance of the AMoD routing.

Rebalancing

The rebalancing procedure part is the handling of idle shuttles (without passengers). One outstanding goal is the service quality, i.e. minimizing the average pickup waiting time for clients. This is achieved by letting idle shuttles approach areas where future requests are expected. The second objective is to minimize the operation costs, which is tackled by minimizing the number of shuttles performing rebalancing trips.

The implemented rebalancing solution in the APEROL simulation context follows a gravitational approach that is based on historic traffic demand schemes for the day. Using these schemes, the areas from where soon to be made requests are known and idle shuttles, without passengers, should approach those areas to be available short-term. The greater the traffic demand from a certain area, the stronger the gravitation pull on the idle shuttles. The metric used to calculate the gravitational impact on a shuttle is measured by its travel time along the shortest route to the traffic demand area. In order to make the assignment of idle shuttles to gravitation centers fairer and more balanced, a specific, massively parallel, Neural Network model, invented by Teuve Kohonen [15] and is known as "self-organizing topological feature map", is used. Here the excitation response of the learning step is functionally associated to the gravitation force. The Neural Network is a scalable, distributed computational scheme that can run on many processing units in parallel (about 9,000 CUDA cores on 2 graphic cards).

3.2.3.3 Vehicle Management

The Vehicle Management mediates between the Fleet Management and the autonomous shuttles in the simulation environment, that constitute the highly flexible transport service for the agents. On the one hand it periodically receives the current states of the shuttles (positions, timetable situation, etc.) and forwards the data to the Fleet Management and its AMoD algorithms. In return, it receives updated and modified shuttle schedules from the AMoD algorithms and sends it to the appropriate shuttles. The data transmission frequency for this is once per simulation second.

3.3 MATSim Setup and Configuration

Apart from the concrete simulation scenarios to investigate, a city specific configuration and calibration is necessary. This refers to the overall scope of the simulation, the exemplary daily traffic demand, the role and scheduling of PT, the definition of service points, and the special handling of signalized intersections. As it is not possible to model the full complexity of the real road transport system, some simplifications must be introduced as well.

3.3.1 Scope of the Simulation Scenario

The MATSim simulation comprises the entire city of Aachen (Germany). Aachen is a medium size city with a total population of approximately 247,000. The center point of the city has the WGS84 coordinates 50°46′32″N and 06°05′01″E. It is located in North Rhine-Westphalia with an area of about 160.85 km².

The number of agents used in the Aachen scenario is about 350,000. Throughout the day they conduct about 750,000 journeys in several different transportation modes. The claim (objective?) is to simulate the complete traffic situation during the day without any down-scaling of agents and road capacities.

The shares of mean of transport for Aachen, in 2017, are listed in Table 3.2 [11, 12].

Table 3.2: Shares of mean of transport for Aachen

Means of transport	Share	Means of transport	Share
On foot	29.8%	On foot	43.0%
Bicycle	11.0%	Bicycle	13.0%
Public Transport	13.0%	Public Transport	10.0%
Individual Transport (passenger)	12.6%	Individual Transport (passenger)	8.0%
Individual Transport (driver)	33.6%	Individual Transport (driver)	26.0%

In Aachen, there are nearly 500,000 routes of less than 5 km travelled each day.

The Public Transport is characterized through [13]:
- Capacity bottlenecks in the public transport network (despite the use of extra-long, double-articulated buses).
- The performance of the current system is limited in several places and can barely be increased while maintaining the quality of the service.
- Cycle density is insufficient. On certain axes there is no uniform timing frequency of service.

- Travel from the outer boundaries to certain sections of the city center are unnecessarily long during off peak times.

Aachen has a relatively high accident rate when compared with other cities with more than 100.000 inhabitants. The five-year average of casualties per 10.000 persons in Aachen is about 45. In comparison in the comparable city "Mülheim an der Ruhr" has about 20 casualties [13].

The MATSim network was generated using Open Street Map data. It consists of about 136.260 directed links and 69.814 nodes. The main criteria for the spatial definition of the overall simulation area included: (1) the available traffic demand information concerning private and public transport, (2) a clear border cut with arterials and motorways, (3) the inclusion of all public transport routes, and (4) in general the coverage of the main transportation paths and traffic dynamics in Aachen. For performance purposes, the network does not contain every single small road. Additionally, these roads are not necessary for the traffic simulation to realistically model the relevant effects of the traffic dynamics. As a result, the generation of the simulation network ignores all road categories that do not support at least cars, public transport, or taxis.

Figure 3.4: The Aachen area concerned (left) and its generated simulation network (right).

3.3.2 Traffic Demand and Agent Plans

In MATSim, the set of agents combined with their daily plans represent the daily transport needs of the real population of the city. The dynamic traffic demand expressed by origin destination relations is, to a certain degree, equivalent to those agent plans. In other words, it is possible to derive the agent plans directly from the origin destination matrices (which contain the number of trips from origins to destinations per time slot). It should be noted, however, that the agent plans do not necessarily have to define the assignment of agents to modes of transport.

APEROL simulation focus is the dynamic traffic demand during normal weekday traffic without any special events (sports, holidays, etc.) as they might impact scenarios. The city of Aachen, very kindly, made individual and public transport (above all bus lines) data available to the research project. The data comes in the form of origin-destination-matrices where traffic flows from every origin to every destination in the network is defined (with unit vehicles per hour). In Aachen, the matrix traffic flow information is even more detailed, as it is broken down by the purposes of the journey (living, working, school, university, services, shopping, leisure, …) and summed up for the whole day. They are not temporally dissolved in form of values per time slot (e.g. for every hour), which is needed for a realistic simulation of the traffic dynamics over the day. Rather, the city council delivered so-named "level information per journey purpose", which is in fact the temporal distribution of the purposes or activities types over the day. The required temporally dissolved traffic demand information was derived by combining and processing the data from these two sources.

Origins and destinations are defined zones that are contiguous sub-areas of the city. All zones form a parqueting of the overall city area (see figure below). In the case of Aachen such zones coincide with the constituencies of Aachen as can be seen in Fig. 3.5. Each district thereby is defined through a collection of street names. For example, district 1 with:

"Annastr. 1-33 u. 2-30, Augustinerbach, Augustinergasse, Augustinerplatz, Büchel, Buchkremerstr., Domhof, Fischmarkt, Friedrich-Wilhelm-Platz, Hans-von-Reutlingen-Gasse, Hartmannstr., Hof, Holzgraben without No. 10, Hühnermarkt, Johannes-Paul-II.-Str, Judengasse 4-6, Karlshof (market),

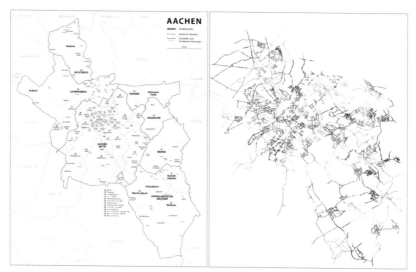

Figure 3.5: Traffic demand data from the city of Aachen: The city voting districts which are taken for origin destination zones for traffic demand (left). And the corresponding parts of the simulation network (right).

Katschhof, Kleinmarschierstr. 4-6, Klosterplatz, Kockerellstr., Körbergasse, Krämerstr., market, Münsterplatz, Pontstr. 1-23 and 2-38, Rethelstr., Ritter-Chorus-Str., Romaneygasse, Rommelsgasse, Schmiedstr., Spitzgässchen, Ursulinerstr."

For simplification, once all journeys (from any origin zone to any destination zone) were assigned to a network link, each zone was assigned an equal distribution of zone links. Consequently, every traveler has a defined start and destination location. During the simulation it is then straightforward for the agents (travelers) to approach the nearest service point or to travel to the service point that is nearest to his destination link.

As for the public transport, it shall be mentioned that the corresponding traffic demand is not defined from bus stop to bus stop but is rather assigned to pairs constituencies similar to the individual transport. All public transport passengers are assumed to walk to the nearest bus stop in order to start their journey.

3.3.3 Public Transport

In Aachen, the complete public transport service is based on busses, many of them double articulated. All these busses are organized by fixed lines, routes and timetables. The network of bus line routes is dense as shown in Fig. 3.6.

Approximately 13% of all journeys, in Aachen, are public transport. There are many opportunities for optimization in the current system. As performance is

Figure 3.6: The bus lines in Aachen.

limited, cycle density can be insufficient and the journey times, from the outskirts to certain sections of the city center, are too long [13]. Some of the simulation scenarios to be presented later address other – more flexible – public transport scenarios (with smaller busses, higher frequency or demand driven routes).

3.3.4 Signalized Intersections

Aachen has about 160 signalized intersections (traffic lights). The APEROL MATSim simulation considers about 140 (see Fig. 3.7).

Depending on the average green split of the traffic light control program the capacities of the approach links vary accordingly. The capacity of a road is measured by vehicles per hour. The capacity has a great impact on the traffic dynamics in the vicinity of the intersection. When the saturation approaches the capacity (in case of heavy traffic), due to the capacity restraint function, the traffic users experience waiting queues at stop lines and longer waits for service. In the APEROL MATSim simulation (1) the travel times are definitely evaluated to find a more suitable configuration pattern for the next iteration, and (2) routing, as an essential part of fleet management, is based on link states (travel times).

The traffic light control (TLC) in Aachen is mostly fixed time, with some exceptions, such as simple overruling bus demand mechanisms, that can be considered as public transport prioritization. This fixed time control data has

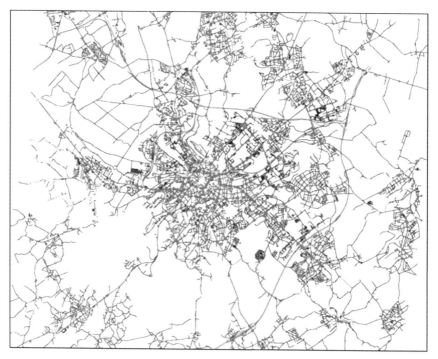

Figure 3.7: Positions of the 140 signalized intersections in the Aachen city simulation.

been used to derive the capacity values of the corresponding network links (like "capacity = green-split * saturation flow"). In a way, such capacities are averaged values. But, as the MATSim simulation is macroscopic on a relatively high level, there is no interest in more details; averages are sufficient.

3.3.5 Service Points (Stops)

Service points for the APEROL mobility scenarios are service or transport mode dependent. As far as today's public transport is concerned, the existing bus stops are used as defined by the ASEAG operator. With respect to the new mobility on demand services with autonomous shuttles many virtual service points were created according to some meaningful criteria:

- The traveler should not walk longer than 150 meters to get to the next service point or to get to the final location from the destination service point.
- The traveler may not cross a main road in order to get to a service point within 150-meter distance.
- The new service points shall not coincide with the existing bus stops.
- The new service points shall be assigned to network nodes (rather than links) and be evenly distributed over the network.

For each service point assigned to any network node an additional (virtual) link will be introduced. The virtual link is very short and connected to this network point. This additional link could be envisioned as a vertical service point queue. As a matter of simplicity, it was not taken into account whether the roads assigned to the service points offer sufficient space for several shuttles to stop there. In justification of this decision, it shall be mentioned that the shuttle scenarios with mobility on demand are future scenarios. Until then the corresponding road construction measures can be carried out.

Figure 3.8 shows the Aachen network with all service points for autonomous shuttles.

3.3.6 City Borders

Besides internal city traffic, there are several relevant traffic flows that cross the city border (going beyond the spatial scope of the simulation). An example for this is the traffic between Aachen and Cologne on the A4 motorway. According to the available traffic demand data, incoming and outgoing traffic is about 30% of the overall traffic. To model these cases, additional (virtual) origin-destination-zones have been defined and are connected with defined network entry and exit links. As a rule, such entry and exit links are motorways or big arterials. The travel times, for the portions of the trip outside the simulation network, are not measured but predefined to fixed values. The fixed values correspond to the distance between the zone and the city network border (the entry and exit links). Such travel times are needed to determine the time point when agents (vehicles) with known travel start times arrive at the entry links of the simulation network.

Figure 3.8: The service points (public transport and autonomous shuttles) in the simulation network Aachen.

3.3.7 Simplifications

The real transport system of such scope (city of Aachen) is of utmost complexity. It comprises plenty of specialties and subtle features that cannot all be addressed by a simulation system because of the following reasons: (1) It is not possible or disproportionately costly to capture all data that is needed to cover such details, (2) the simulation model would be overloaded (i.e. hard to configure and calibrate), and (3) the overall simulation performance would be unacceptable.. Therefore, and because we want to simulate the road traffic of the entire city, simplifications were introduced in APEROL. The most important ones to mention here are:

- Pedestrians and bikers are not modelled. Consequently, footpaths of agents at the beginning or the end of their journey are not considered. Transport in our context always means to use a vehicle (autonomous car, bus, …).
- Multimodal trips are not modelled. In the APEROL simulation world, nobody can for instance use a bus for the first part of the journey and then change to an autonomous vehicle.
- Train transport is not modelled, as its impact on the load of the Aachen city network is neglectable. In addition, journeys to and from railway stations are already included in the individual journeys of the other means of transport.

- The process of "negotiating" a transport service for AS is not included. Just one offer per request will be generated by the AMoD-algorithm and immediately acknowledged and used. Any special request conditions, such as late bookings or requests which are impossible to fulfil, will not be dealt with.
- Trip cancellations are not considered.
- With regard to transport demand and transport plans of a population, only one single representative day is considered. The means of transport and the starting times might change from iteration to iteration.
- Special events (sports, musical), accidents, road closures and obstacles (bus breakdowns, delivery traffic, waste disposal) are not considered.
- Pedestrians as road users are not considered. Although current statistics show that about 30% of the distances travelled in Aachen are on foot, pedestrians have their own transport infrastructure that are mutually exclusive to the roads studied in terms of capacity.
- AMoD shuttles are only booked promptly on demand and not a long time in advance. The fleet management does not need to handle bookings with travel starts that are several hours in the future. Moreover, already booked shuttle trips are always fixed and will not be changed or canceled.

3.3.8 Scoring Functions and Replanning Strategies

MATSim offers a co-evolutionary algorithm to adapt agent plans over many iterations and eventually arrive at an equilibrium. The equilibrium can be regarded as an optimization within a certain framework of constraints, where the agents cannot further improve their plans unilaterally.

The co-evolutionary algorithm is subdivided into two steps: (1) After each simulation run or iteration, the actual performance of every agent's plan is calculated by scoring functions. (2) On the basis of the scoring values, and other state variables, some agents modify their collection of plans by copying and mutating single plans. Four planning dimensions are usually considered for MATSim: departure time, route, mode and destination. In the context of the APEROL simulation, only the transportation mode shall be influenced by co-evolution.

Scoring

To define the scoring functions a cooperation with the Human-Computer Interaction Center (HCIC) of RWTH Aachen University was established. The challenge was to identify all relevant decision criteria of the users (attributes of the mobility offer) and to define the gradations of the attributes in such a way that users understand them in a survey, but at the same time they could be translated into simulation parameters. The following criteria were declared as relevant:

1. Travel time (relative to minimum transport duration)
2. Walking distance
3. Costs/kilometer

4. number of passengers
5. Possible delays at destination
6. Own means of transport
7. Waiting time (relative to desired departure)
8. Number of transfers

For all those criteria up to five meaningful levels were defined. Example for the first criteria (travel time): $L1$ = shortest possible time, $L2$ = 5 minutes longer, $L3$ = 10 minutes longer, $L4$ = 15 minutes longer, $L5$ = 20 minutes longer.

Moreover, meaningful scenarios were defined by combining survey results of 200-300 participants. A conjoint analysis was conducted on the interview data. The scenarios did not flow directly into the conjoint, but rather provided the framework for them. Finally, from the survey results the scoring function parametrization were derived.

Replanning

As for replanning, the standard functionality was used that comes with MATSim. No special replanning strategy were implemented.

3.4 Simulation Experiments

One of the outstanding APEROL goals is the investigation of mobility solutions where autonomous, on-demand shuttle services play a more or less decisive role. In this context, a main work package is an agent-based simulation all traffic in Aachen and based on an extended version of MATSim that interacts with a Mobility Center. In this context, the simulation environment must be regarded as a test bed for fleet management algorithms and optimization procedures (AMoD algorithms) and to evaluate different future traffic scenarios with different distributions of traffic modes.

In order to investigate possible alternatives of mobility systems and services, some scenarios were defined. Such scenarios represent certain phases on a mobility roadmap, thereby also showing different automation degrees. The simulation results of those scenarios deliver valuable information about performance, costs, user acceptance and fleet sizes.

It should be mentioned that the simulation also includes the behavior of the road users/travelers. This is because by means of specific MATSim scoring functions and re-planning rules the influence of psychological variables is part of the modelling.

3.4.1 Scenario Definition

The simulation-based mobility investigation in the APEROL context follows a scenario approach, which addresses the most probable and relevant aspects of future traffic systems for urban areas. Such scenarios mainly define possible transportation mode patterns. The idea is to start with the traffic situation as it is today in Aachen, and to experiment with conceivable future scenarios. The future scenarios are characterized through a more flexible form of public transport or a bigger share of on-demand (autonomous) shuttles. The three scenarios are shown in Table 3.3 (APT = Autonomous Public Transport):

Table 3.3: The three Aachen scenarios that has been investigated in the APEROL simulation environment

No	Name	Description	Subjects of Interest
1	Today	• IT, PT shares Aachen 2020	• Is it possible to model the Aachen traffic as it is today realistically?
2	More flexible APT	• Smaller IT share than Aachen 2020 • Existing PT stops • Smaller buses, higher frequency • Semi-flexible lines • Defined user perception and behavior (replanning)	• What transportation mode drift can be detected? • How would the PT / APT system scale if more and more private car travelers change to PT /APT? • What would be the consequences for the overall traffic in Aachen in terms of traffic load and waiting times?
3	AMoD Shuttles	• Complete traffic demand in Aachen is served by AMoD shuttles of different sizes	• How many shuttles are needed to guarantee acceptable waiting times? • What would be the consequences for the overall traffic in Aachen in terms of traffic load and travel times?

The purpose of the first scenario "Today" is to calibrate the simulation environment with its configuration and databases. The methodology to prove this is to compare a workday scenario in Aachen using information from MATSim with TomTom® (TomTom Traffic™ API). A good match is a strong indication that the simulation is close to reality. Therefore, the main objective is to see similar traffic load network patterns and their temporal evolution in MATSim and TomTom Traffic™. Of particular interest here are the patterns of heavy traffic. Note that for the simulation conduct, only one iteration is necessary as the transportation mode shares are fixed at the current modal spilt for Aachen.

In the second scenario "More flexible APT", the APT service shall be denser and more flexible as it is today in Aachen. More flexible means smaller (autonomous) buses, higher frequency, and variation of the routes within certain tight limits. This scenario would investigate how many private car travelers would change to the Public Transport if the APT offering is more attractive. This drift from IT to PT is expected to be caused by scoring and replanning (the perception and behavior of the travelers) after several iterations. To the primary goal is to acquire some indication of (1) how strong the drift is, (2) impacts to the size of the

PT fleet and (3) what the consequences are for the overall traffic in Aachen. The main goal is to determine the added value of such measures, which might be a next migration step towards a more flexible and sustainable transport system. Note that for the simulation conduct only one iteration is necessary, as the transportation mode shares are fixed.

The last scenario "AMoD Shuttles", assumes that all daily passenger transport in Aachen is served only by AMoD shuttles of different sizes. This means implicitly that there are no private cars anymore, and the Public Transport has reached its ultimate form in terms of flexibility and convenience. Of utmost importance here is the question for the size of the necessary shuttle fleet. And also, what would this mean for the shuttle traffic in Aachen (traffic flows, travel times, overall load, …)?

As for the fleet sizing of APT and shuttles, the approach is to test several sizes in the simulation. The evaluation of the experiments determines averages, standard deviations and maximum values of (1) passenger waiting times and (2) travel times. The optimal fleet size obviously depends on definitions like acceptable waiting and travel time statistics (measuring the passenger satisfaction), acceptable traffic conditions (the objectives of the road operator) and also operation costs (which are not considered here). Combining such assessment criteria with the functional interrelation between fleet sizes and waiting times will help determine the optimal fleet size.

3.4.2 Simulation Results

The three proposed scenarios were investigated by means of the APEROL simulation environment that consists of the traffic simulation part MATSim and the APEROL Mobility Center (incl.Fleet Management and its related AMoD algorithms). The simulation runs are based on the current traffic demand (in terms of number of trips) of Aachen.

Regarding the scenarios that required several iterations, 15 iterations have been carried out.

At the time of writing this chapter, the simulation experiments are still ongoing (the APEROL project ends in December 2020). For now, the simulation is still subject to certain restrictions and the results presented here are initial findings. The findings are already relevant and reliable, but more comprehensive simulation experiments are planned for the upcoming months (to be published in subsequent papers, particularly the results of scenario 2 that are not yet available).

Scenario 1

The results of scenario 1 demonstrate that by means of the configuration and calibration of the extended MATSim simulation it is possible to model the current traffic of Aachen with its current modal split realistically. For this scenario, an average workday was chosen, and several simulation iterations were carried out in order to optimize the route pattern of the simulation. Then, the travel times of the network links were compared with corresponding historical travel times from the TomTom® service "TomTom Traffic Stats" [17]. As can be seen in the chart below, the travel time classes coincide to a great extend with the historical ones received from TomTom®. Let L be the set of all network links, h be a time slice of one hour, and L_h^* the set of links l_k^h, for whose mean simulation travel time $tt\_sim_k^h$ the following holds: $tt\_sim_k^h \in \left[\frac{1}{2} tt\_tom_k^h, 2tt\_tom_k^h \right]$, with $tt\_tom_k^h$ being the TomTom mean travel time of link k for time slice h. For example, as for the rush hour at 8 am the following ratio was found: $\left| L_{8am}^* \right| \cong 0.895 |L|$. In addition to the travel time ratio values, it was found that the traffic congestion patterns in the network, derived from TomTom and simulation values, were very similar, which is indication for a very similar traffic dynamic also.

Scenario 3

In the "AMoD Shuttles" scenario, the entire passenger transport is realized by autonomous shuttles. Several simulations were conducted with different shuttle fleet sizes (but fixed ratio of shuttle sizes) for the traffic demand of Aachen. For the time being, because of performance considerations, the 24-hour simulation is down-scaled to 10% of the actual traffic demand. The next planned version of the APEROL simulation environment, which is under development, is going to overcome such scaling limitations.

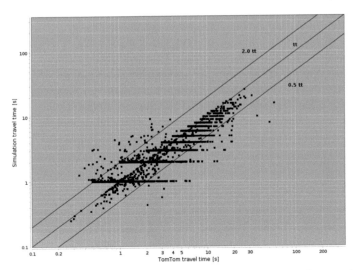

Figure 3.9: This chart shows the correlation of TomTom and simulation travel times for all network links. As can be seen, the majority of the links are located within the diagonal slices. Note that the majority of the simulation travel times are rounded to integers (with the exception of those links without any traffic where the travel time is estimated).

The goal of the simulation experiments was to define the optimal size offering an acceptable service to the clients (i.e. pickup waiting time smaller than a given threshold). Another question of interest was the saturation of the road traffic system through the shuttle fleet (above all travel times).

The simulation outcomes, which are depicted as correlations between mean pickup waiting times, unserved trips and sizes of autonomous on-demand shuttle fleets, are collected in Table 3.4 (SD = standard deviation):

Table 3.4: The simulation outcomes for different fleet sizes in form of pickup waiting times and unserved trips

Fleet Size (# shuttles)	Pickup Waiting Time [s]			Unserved Trips	Shuttle Requests
	Mean	SD	Max		
10000	352	352	3600	4980	74
12500	68	86	3600	30	60
15000	32	48	3600	10	50
17500	20	32	1115	0	43
20000	14	24	778	0	37
25000	9	15	666	0	30
30000	5	9	558	0	25

In the case of a trip being unserved at all, the waiting time was set to 3600 seconds. Both the table (Table 3.4) above and the following chart (Fig. 3.10) establish the rule that the smaller the fleet size, the greater the passenger pickup waiting times and the higher the amount of unserved trips.

It is assumed that a reasonable and acceptable mean pickup waiting time is less than 3 minutes with a standard deviation of 2 minutes. From this, it can be derived that a fleet size of about 20.000 shuttles (with equal share of 7 and 15 seats) would be capable to handle all trips in Aachen throughout the day, which

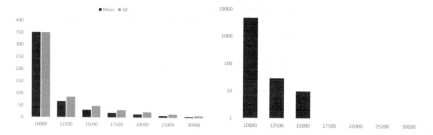

Figure 3.10: Left: The x-axis of this chart shows the shuttle fleet sizes. The corresponding pickup waiting time values (per fleet size) are drawn (mean and standard deviation). Right: The number of unserved trips per fleet size.

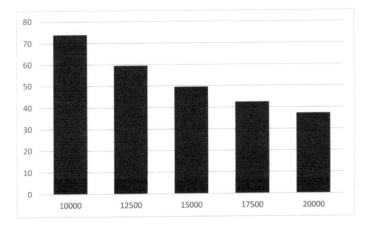

Figure 3.11: The average number of requests per day the shuttles serve depending on the fleet size.

today are assigned to individual and public transport. It is also obvious – although not shown here explicitly – that such a shuttle fleet would yield a much smaller traffic load in the Aachen network.

3.5 Conclusion

In this APEROL research, a MATSim-based simulation environment was introduced. This specific environment is able to model urban road transportation systems, which either include (1) an enhanced and more flexible Public Transport with smaller and more numerous vehicles, or (2) integrated fleets of autonomous shuttles that realize a mobility-on-demand system (AMoD). It was shown in detail how the available traffic demand information for the city of Aachen was pre-processed in order to create the corresponding simulation configuration. Moreover, AMoD specific functional extensions, which are implemented and integrated in MATSim, were specified. After the calibration of the environment, several possible AMoD-based scenarios were investigated.

In terms of AMoD services, it could be demonstrated that a fleet of about 20,000 shuttles would be sufficient to offer an acceptable service performance (in terms of pickup waiting times) for the city of Aachen with more than one million trips per day,.

Future research activities will be devoted to investigating an extended and more flexible Public Transport system with more and smaller busses with varying line routes within certain boundaries. Furthermore, the simulation of realistic AMoD services will be enhanced and amended. Special focus will be put on monetary cost calculation and optimization.

References

1. Pavone, M. 2015. Autonomous Mobility-on-Demand Systems for Future Urban Mobility. Autonomes Fahren. Berlin, Heidelberg, Springer Vieweg.
2. Hörl, S. 2017. Agent-based simulation of autonomous taxi services with dynamic demand responses. *Procedia Computer Science*, Bd. 109, 899–904.
3. Wang, B., S.A.O. Medina and P. Fourie. 2018. Simulation of autonomous transit on demand for fleet size and deployment strategy optimization. *Procedia Computer Science*, 130, 797–802. https://doi.org/10.1016/j.procs.2018.04.1.
4. Maxime Guériau and Ivana Dusparic. 2018. SAMoD: Shared Autonomous Mobility-on-Demand using Decentralized Reinforcement Learning. *In*: IEEE International Conference on Intelligent Transportation Systems (ITSC).
5. Alonso-Mora, J., S. Samaranayake, A. Wallar, E. Frazzoli and D. Rus. 2017. On-demand high-capacity ride-sharing via dynamic trip-vehicle assignment. Proceedings of the National Academy of Sciences, pp. 462–467. doi:114. 201611675. 10.1073/pnas.1.
6. Martinez, L.M. and J.M. Viegas. 2017. Assessing the impacts of deploying a shared self-driving urban mobility system: An agent-based model applied to the city of Lisbon, Portugal. *International Journal of Transportation Science and Technology*. 6(1), 13–27
7. Bischoff, J. and M. Maciejewski. 2016. Simulation of city-wide replacement of private cars with autonomous taxis in Berlin. *Procedia Computer Science*, 83, 237–244.
8. Horni, A., K. Nagel and K. Axhausen. 2016. The Multi-Agent Transport MATSim. London: Ubiquity Press.
9. Hörl, S., C. Ruch, F. Becker, E. Frazzoli and K.W. Axhausen. 2018. Fleet control algorithms for automated mobility: A simulation assessment for Zurich. TRB Annual Meeting Online.
10. Horni, A., K. Nagel and K.W. Axhausen 2015. The Multi-Agent Transport Simulation MATSim. London: Ubiquity.
11. Verkehrs-Entwicklungs-Planung Aachen. 2018. 7. Lenkungsgruppe. VEP Aachen, 7. Lenkungsgruppe, 29. Nov. 2018.
12. Aachen.de. 2017. "Zähldaten" 2017. http://www.aachen.de/de/stadt_buerger/verkehr_strasse/ver kehrskonzepte/Radverkehr/zaehldaten/index.html
13. Mobilitätskonzept Aachen 2020. Stand: Oktober 2013, https://nrw.vcd.org/fileadmin/user_upload/NRW/Verbaende/Aachen-Dueren/MOBILITAETSKONZEPT_AACHEN_2020_131012.pdf
14. MATSim Multi-Agent Transport Simulation. www.matsim.org
15. Kohonen, T. 1982. Self-organized formation of topologically correct feature maps. *Biol Cybern*, 43, 59–69.
16. Research project APEROL home page, https://www.autonomousshuttle.de/
17. Access to Traffic Stats data from TomTom International BV, https://developer.tomtom.com/traffic-stats

Artificial Intelligence for Fleets of Autonomous Vehicles: Desired Requirements and Solution Approaches

**Matthias Dziubany[1]*, Sam Kopp[1], Lars Creutz[1],
Jens Schneider[1], Anke Schmeink[2] and Guido Dartmann[1]**

[1] Institute for Software Systems (ISS), Trier University of Applied Sciences,
  Birkenfeld, Germany
[2] ISEK Research and Teaching Area, RWTH Aachen University, Aachen, Germany

4.1 Introduction

With the appearance of autonomous vehicles, a substantial change in transportation systems is to be expected. The progress in communication and computing power already indicates a great change as mobility on demand and ride sharing gets practicable and popular.

Instead of trying to survey all recent transportation system proposals, concerning fleets of autonomous vehicles, such as [1] and [2], this chapter begins with a collection of main desirable requirements from a user's perspective, depicted in Table 4.1. It then concludes with optimization strategies from literature that

Table 4.1: Collection of desirable requirements

Requirements
Demand-responsive
Single and shared modes
Heterogeneous vehicles
User co-passenger preferences
Time window consideration
Dynamic
Congestion avoidance

*Corresponding author: m.dziubany@umwelt-campus.de

fulfill some of these requirements. Surely, this collection of requirements is not complete. As it was set up by experts from different domains like mathematics, computer science, simulation, economics and acceptance research, it is however reasonable from many points of view.

Unfortunately, there is no literature example that covers all requirements. Especially since user preferences have been neglected for a long time, but a new trend can be recognized. For this reason, mainly transportation systems, that take the individual preferences of citizens into account, are reviewed [27, 28]. In particular, the idea of graph-constrained coalition formation [26] is adapted to the collected requirements. Besides a step towards individual transportation, more deterministic traffic values with the appearance of autonomous vehicles can be expected, since a lot of data is recorded and ideally, most vehicles are routed by one entity. Therefore, especially solutions concerning the impact of their own plans on the future traffic conditions are introduced in Section 4.5.

Finally, it is concluded that multicommodity-coupled flows in time-expanded networks are satisfactory to formulate the optimization problem based on our requirements, and the corresponding integer program (IP) is stated and explained. Due to the current lack of computing power, it is not possible to solve this big problem optimally. In conclusion, distributed algorithms from literature are reviewed.

The desired requirements depicted in Table 4.1 are explained in Section 2. In the subsequent section, the requirements are transitioned into a user interface. After that, the focus is set on fulfilling user preferences in Section 4. Here, the idea of graph-constrained coalition formation is explained and adapted to the collected requirements. As this solution concept neglects traffic control, Section 5 investigates congestion avoidance. Finally, in Section 6 the optimization problem fulfilling all requirements is formulated as an integer program using the idea of multicommodity-coupled flows in time-expanded networks. After reviewing distributed algorithms, which scale well with the number of requests and vehicles, in Section 7, the chapter ends with a conclusion and outlook in Section 8.

4.2 Desired Requirements for Future Transportation Systems

At present, most citizens have largely to adapt to the transportation system (fixed stops, fixed times, expensive taxis). With the progress in communication, mobility on demand systems like Uber [3], Lyft [4], DiDi [5], Careen [6] and MOIA [7] become more and more popular. These systems lessen the barriers for younger, elder or disabled people as it is no longer necessary to own or be able to drive a car, to experience an individual transportation solution [11] [18]. In inner west Sydney the demand responsive transit service Bridj is applied by transit systems to fill the coverage gaps considering the broader public transport network [8]. Certainly, the appearance of autonomous vehicles will also have a positive impact on demand-responsive transportation, since no drivers are required. So far, autonomous shuttles used for public transportation drive only on fixed routes like the five bus lines in Monheim [9]. In order to adapt to different transportation needs, the flexibility in their route has to be increased. Since a ride can no longer fail or be postponed due to issues with the vehicle operator, an autonomous system is therefore more reliable. At best, the transportation system optimally adapts to the demands and needs of all citizens. In order to operate this kind of system, the problem should be modeled first. Unfortunately, it is hard to identify all desired requirements before having the system in operation. A group of experts from different domains made a first attempt. Their main desired requirements are explained below.

4.2.1 Demand-responsive

Buses with fixed routes and schedules can offer a great throughput and are easy to manage. However, the systems often waste a lot of resources when there is no frequent demand for transportation, which is shown by buses or trains operating, although they are empty [10]. To get rid of fixed schedules is a big challenge for ride-sharing services of all kinds [14], but providing demand-driven rides can be more appealing to people, as it provides flexibility and spontaneity [22]. Also, in many situations, a demand-responsive transportation system is able to serve customers faster and with less vehicles by driving variable routes. In general, this approach is more efficient and also more convenient for users, because they can choose door-2-door transportation, which is particularly attractive for elderly and handicapped people. Furthermore, a study indicates that the overall acceptance of autonomous vehicles is higher, when a car is directly available upon request [17]. It is thus concluded, that the future transportation system should be demand-responsive. This requirement however, does not exclude big buses on popular routes, if a fleet of heterogeneous vehicles with a different number of seats is considered.

4.2.2 Heterogeneous Vehicles

Most vehicle routing literature considers heterogeneous vehicles only in the transportation of handicapped people. They take care if there is a spot for a wheelchair or certain equipment on board [25]. But commonly, every customer has vehicle preferences. Some people may prefer to drive in a small bus instead of a big one or even wish a luxury vehicle, so these factors have to be considered in an autonomous fleet, too. As users are no longer busy with operating the car, they find themselves in a public transport situation, where they expect an experience that is at least as convenient or even more convenient than using current public transportation [17]. Surveys show, that people also like to do activities, similar to those in public transportation. In related studies in Germany, people were asked for their favorite activities to do in an autonomous car and mostly answered with relaxing/sleeping, looking out the window, listening to music, talking to other passengers, texting/being on the phone or working [21, 23]. These results correlate with similar surveys in other countries, which add, that some users also prefer to entertain themselves by reading or watching TV [15]. Offering all these amenities in an autonomous transportation system are seen as an attractive feature and a great advantage over traditional cars [11, 16, 20], and are even likely to help maintaining enjoyment and willingness to use autonomous vehicles in general [19]. Therefore, an autonomous fleet has to consist of vehicles that differ in size and facilities. There need to be bigger buses for people that like to chat with co-passengers as well as smaller cars, with carefully chosen occupants, allowing them to work or relax. Some might even offer infotainment displays or special entertainment systems for games. From the system view, a heterogeneous fleet has many advantages, too. For instance, a double-decker bus is able to carry and serve many people at once, while small vehicles are much more efficient in handling a single customer.

4.2.3 Single and Shared Mode

"For commuters, major rideshare benefits include travel time savings, cost savings (namely fuel and parking) and increased mode choices." [14]. To fulfill these requirements, the system has to offer different options, regarding the amount of people, a car is shared with. Privacy is crucial for many people, when it comes to traveling [17]. That is why driving alone can be an important requirement for some users. The transportation system has to consider this preference and provide appropriate privacy features. These can range from a more convenient car interior or increased space per passenger [17], to offering single rides in an autonomous vehicle. If the user is able to afford occupying a car on his own, he should be given this opportunity to enhance the system's attractiveness.

Within the fleet, single mode requests should not be assigned exclusively to taxi systems, since the destination of a single mode drive can be on the way to shared mode riders, which do not have the same starting and destination location. Nevertheless, shared mode should be the default ride mode, as it saves fuel and costs.

4.2.4 User Co-passenger Preferences

With a heterogeneous fleet and the possibility to choose from single or shared rides, many user preferences can be satisfied. For users that opt for riding in shared mode, there are still genuine concerns about the passengers, they share a vehicle with. People often express feelings of discomfort when they drive with complete strangers, also known as "stranger danger" [14], which are especially strong at night [18]. In conclusion, users could feel safer at night with people they know or parents might want their children to share rides only with familiar passengers. Besides privacy, that has been mentioned before, safety and trust are also important factors for ride sharing [11, 14, 15, 17]. So, the system should allow driving with friends, respectively people that know each other. Thus, fearful users can be satisfied and it boosts the overall willingness to use a ride sharing system in the first place, as well as the overall travel satisfaction [13].

4.2.5 Time Window Consideration

With time windows, the waiting and driving time of a customer can be restricted. Business people on their way to work, for example, choose small time windows in order to lose as less time as possible for transportation [20]. Other customers are fine with larger time windows, when they are on a leisure trip [20] or if their ride is cheaper then. Generally, studies indicate that waiting and driving times are an important factor for all different kinds of people, because shared rides have to compete with the time flexibility and efficiency of owning a personal vehicle [12]. Since most requests are expected to have small time windows and come in spontaneous and not hours or days before, the transportation system also has to be dynamic.

4.2.6 Dynamic

Requests to the transportation system can either be classified as static or dynamic [24]. In the static case, all requests are received up to a certain point of time before the customer wants to actually take the ride. The exact time point is variable and can spread from hours to days before the trip. The main advantage of these requests is, that the system can optimize and bundle tours, happening at the same time, since it simply knows all inquiries up-front. Unfortunately, not all transportation demands occur long before a ride, because people like to travel spontaneously and appreciate a system, that can handle this requirement [22]. We consider these requests to be dynamic, as they can appear suddenly and with an instant transportation desire. The system has to be able to include these sudden demands and fulfill them quickly, by checking if there is a vehicle nearby or already on a trip, that can serve the new customer. If rescheduling a driving vehicle, and thereby pooling users to the same ride, is the most efficient option, the car's routes have to be optimized and adjusted dynamically. In conclusion, the system needs to process requests that were not known prior to initially planning and starting tours. Considering this, there are probably some requests that cannot be served within the favored time frame and therefore have to be rejected in the dynamic case. The user then has to choose different settings for his trip and try again.

4.2.7 Congestion Avoidance

The general knowledge of and control over all or many vehicle routes do not only enable traffic prediction, but also congestion avoidance. This can be achieved by intelligent vehicle routing, in order to avoid bottlenecks on roads that are usually prone to congestion [11]. Also, with more people using shared rides, the total amount of cars can be decreased, furthermore reducing CO_2-emissions as a side-effect. This potential should definitely be utilized in future transportation, as it can be of high value for the entire traffic system. It should be added, that fewer vehicles also lower the need for parking space [11, 17], particularly in urban areas. This is perceived as a great advantage by users, too [18].

4.2.8 Further Requirements

The list of requirements, presented in this chapter, tries to summarize the main concerns, but is only a small fraction of expectations, that people have towards an autonomous transportation system. It has to be mentioned, that one of the top priorities for accepting this kind of transport are solid security measures. Multiple surveys show, that people are extraordinarily concerned about handing off the vehicle-control to a computer and still being safe in a self-driving car [11, 14, 15, 17]. However, this requirement can only be partly satisfied by the fleet-managing system, described here, although it naturally calculates the best and most secure routes. It is mainly a task, that the individual car has to take care of and hence is not listed in this chapter.

People also name reliability as an important factor for ride-sharing situations [17]. While traditional ride sharing is dependant on a driver to operate the vehicle and stick to the time and route of the arranged trip, an autonomous system is more flexible. Since the routing and schedules are highly dynamic, a confirmed ride can always be rescheduled, for example when the planned car or tour fails due to changing traffic conditions, technical failures or accidents. The user might have to accept a longer waiting or driving time, but can still be offered another route or another car. Furthermore, a transportation without drivers is available at any time, unlike taxis or traditional ride sharing services.

4.3 A Request Interface of the Desired Transportation System

In this section, the formerly listed requirements are transitioned into a user interface design for a fictional app, that allows a customer to interact easily with the transportation system. The following pictures depict how a ride with an autonomous vehicle can be booked by a user and what possibilities and configurations for a trip can be offered. The shown process only deals with this particular use-case and is just one part of a potential interface that would be needed to communicate with a complex system like this. The focus was set on a booking process, that is easy to understand and quickly to complete, because a disproportionately high effort to order a vehicle in the first place is considered a great handicap [14]. However, the presented design did not undergo a complete and proper design process, and can therefore solely be seen as a first idea towards a sophisticated user interface.

Figure 4.1 shows the first step of the booking process. The user is supposed to enter a location for pickup and drop-off at the top. The buttons next to the fields

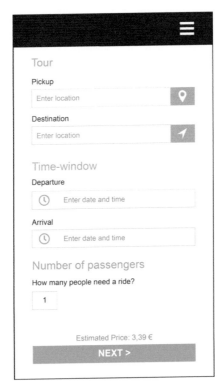

Figure 4.1: App interface to choose destination, time and group size for the ride.

open a map and the user can pin a location. For entering a pickup location the customer can also enable the device's GPS on the map window and read the real location as input. Next, there will be a time window needed for the ride. The user has three options:

1. A desired departure time can be entered and the arrival time is left blank. This is useful for spontaneous trips, which want to be started as soon as possible but do not care about the exact time of arrival.

2. A desired arrival time can be entered and the departure time is left blank. A customer might want to arrive at or before a certain time, because he has to meet a deadline. This could be any ride to a fixed appointment, either business-related or leisure trips.

3. The user may enter both, a desired time of arrival and departure, when it's crucial to journey within a particular time frame. However, this option and the option before leads to further constraints in the underlying optimization problem, to find an appropriate ride. There might not always be an offer (solution), that can fulfill these constraints, so the request needs to be rejected at the end of the booking process. The customer then has to choose a different time interval.

The corresponding input fields are thus partly optional, but there must either be an arrival or departure time. The text inputs could also open a date/time picker upon click for an easier handling of time formats.

At the bottom, the user is asked to enter whether he plans to ride alone or in a group of several people. The estimated price above the "NEXT" button gives a first approximation of the trip's final price. It can be found on each of the following views and changes, depending on the settings the user chooses, in order to give an instant feedback.

Upon clicking "NEXT", the user is lead to the following view, which is shown in Fig. 4.2. On this page, the customer can choose between different types of vehicles, to account for the different convenience and privacy preferences, as explained in the previous section. Depending on the selected vehicle, the facilities are dynamically displayed, because these properties can differ between types. The list of items is only a suggestion of what might be possible to offer in this scenario. The customer can select multiple options for his ride. For instance, a space for luggage (e.g. a bag or case) and a window seat is requested (it is assumed that regular cars can also offer a middle seat). The estimated price is now higher than before, because the order contains these extras.

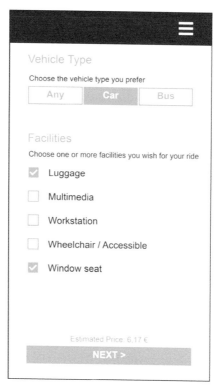

Figure 4.2: App interface to choose the preferred type of vehicle and facilities.

Figure 4.3 illustrates the next step, that asks for the mode to travel. A customer can ask for a single ride in a vehicle or choose the shared mode, which should be selected by default. If a shared ride is chosen, the user can also opt for a preference, regarding his co-passengers. Thus, people that like to work during the journey can travel with other business people and people that do not want to drive with strangers can select "Friends". Friends can be considered people that have already shared a ride together and liked it or people that know each other in real life. For this feature, it is necessary to have an appropriate user management with some kind of social-network structure and the ability to add someone as a friend. The complete order can be finalized and sent to the system by clicking on "GET OFFER".

The system then takes all parameters and calculates a possible ride, which fits the given inputs as best as possible. After completion, a response is sent back to the user and the app displays the screen that can be seen in Fig. 4.4. The customer receives complete information about the offered ride and is finally given a real price for the journey. The offer may slightly vary from the original request, when there is no trip available, that fits every single parameter. In the shown case, the

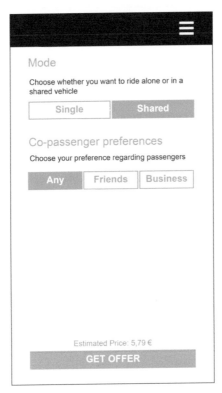

Figure 4.3: App interface to choose the preferred ride mode and co-passenger preferences.

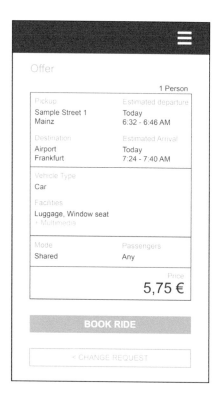

Figure 4.4: App interface to show the customer an offer for a ride.

vehicle has the requested facilities, but is also equipped with a multimedia system for movies, games, etc. In general, any of the ride's properties could vary from the request, if the system still considers the offer suitable. Of course, there are more and less important constraints. For example, diverging from the requested location or time window might not be allowed. A separation between "must-have" and "nice-to-have" requirements could solve this problem, but is not considered in this chapter.

    If the user is satisfied with the offered trip, despite any potential variations, he can book the ride. Otherwise, he can go back and change the request to get another offer. As stated previously, requests that cannot be fulfilled at all may occur and the user has to change his order parameters right away.

4.4 User-centered Transportation Systems

Nowadays, most people drive with their own car instead of using public transportation, with all well-known consequences, such as crowded streets and polluted air. In order to overcome these downsides, the main goal of a shared transportation system should be user satisfaction. Recently, a change from system optimization to individual preference optimization is also noticeable in the transportation-system literature. Instead of just minimizing costs or maximizing overall system profits, the objective becomes more user-centered.

In [26] for example, the overall user satisfaction is maximized, where every user can have an individual objective function, containing attributes like driving time, time window violation and number of co-passengers. The authors in [27] defined this function by machine learning, instead of giving weights to each of their attributes. They considered travel cost, travel time, number of co-passengers, users seat, working status and user demographic information in their machine learning model. For single-mode rides, even personal routes are computed, concerning individual route preferences like barely car accidents, scenic, straight, flat and low air pollution [28]. Especially people, who do not trust autonomous vehicles that much, would appreciate a route with less critical driving scenarios like intersections, road narrowings or highways.

In the shared-mode literature, co-passenger preferences are recently considered. The authors in [29] suggest to maximize an utility function, containing an operational value and a user value. In the user value, the age, gender and rating of users are considered. A cost of discomfort between each pair of customers is assumed to be given in [30]. The work in [31] considers a ride-sharing social network for the passenger matching in taxi sharing. People who know each other are allowed to have greater detours than strangers. The ride-sharing social network is continuously updated, when people share a ride with each other multiple times. In the deployment of autonomous vehicles, the stranger danger is expected to be even greater, since there is no more driver, who is trusted by many people [32].

Instead of incorporating user preferences in the objective function, there are literature examples that guarantee a user preference fulfillment by constraints. This exact fulfillment is especially needed in the transportation of handicapped people [25]. For example, there has to be a place for a wheelchair or the vehicle needs to be specially equipped.

In the following, the idea of graph-constrained coalition formation [26] is explained and adapted to our requirements, since this optimization technique allows to give each user an individual objective function. Further, hard constraints like vehicle choice and co-passenger restriction can be incorporated.

4.4.1 Routing by Graph-constrained Coalition Formation

With the application of graph-constrained coalition formation to the ride-sharing problem, Bistaffa et al. captured co-passenger preferences in their assignment and routing optimization [26]. There, a twitter graph was used to restrict the formation of ride-sharing groups. By contracting and coloring the edges of the twitter graph, an efficient search tree was created and used for optimization. Their procedure guaranteed, that each person was known by at least one other person in the vehicle. In order to capture vehicle choices and stricter co-passenger restrictions, this idea can be extended by setting red (dotted) edges, that forbid contractions from the beginning. In this case, coalitions are only feasible, if they contain no red (dotted) edges and are connected by green (dashed) ones.

For example the coalitions $\{A_1, K_1\}$ and $\{A_3, K_2, K_3, K_4\}$ with vehicles $A_i, i = 1, \ldots, 3$ and customer $K_j, j = 1, \ldots, 4$ are feasible in the graph of Fig. 4.5. In contrast, the coalition $\{A_1, K_1, K_2\}$ is not feasible, since there is a dotted red edge between customer K_1 and K_2.

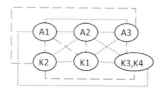

Figure 4.5: Constraint graph, where dashed green edges respectively dotted red edges allow respectively forbid coalition formations.

The search tree can still be created like in [26], since they use the idea of red (dotted) edges, to forbid same coalition-structure generations in the parallel branches.

Below, we demonstrate, what can be achieved by so-called 2-coloured-graph-constrained coalition formation with small examples.

1. Customer K_1 can choose, which vehicles $A_1, \ldots, A_3$ he wants to drive with. In Fig. 4.6 for example, customer K_1 does only want to drive with A_1 and A_2.

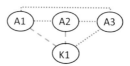

Figure 4.6: Constraint graph ensuring that customer K_1 does not form a coalition with vehicle A_3.

2. Customers can decide, if they want to drive alone, only with known persons or with arbitrary ones. Further, customers can restrict to co-passenger-respecting silence. In Fig. 4.7, customer K_1 wants to drive alone, while customer K_2 and K_3 allow each other as co-passenger.

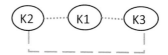

Figure 4.7: Constraint graph ensuring that customer K_1 does not form a coalition with customer K_2 and K_3. In contrast the dashed green edge between K_2 and K_3 allows them to be in the same coalition.

3. Customers can fix each other as co-passenger. In Fig. 4.8 customer K_3 and K_4 want to ride with each other definitely, while they also allow K_2 in the same vehicle.

Figure 4.8: Constraint graph in which customer K_3 and K_4 already formed a coalition.

Some coalition-formation constraints cannot be captured by the graph and have to be included in the creation of the search tree. For example, each coalition with cardinality greater one has to contain one vehicle and the coalition is not allowed to be greater than the number of seats plus one of the assigned vehicle. These constraints can be included in the search process as follows.

1. By contracting only nodes that contain a vehicle, it is guaranteed that each coalition contains one. Figure 4.9 gives an example of an allowed contraction and point out an edge, that is not allowed to be contracted, since the connected coalitions do not contain a vehicle.

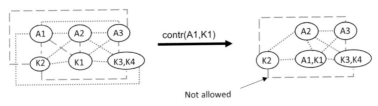

Figure 4.9: Allowed contraction contr(A_1, K_1) and a marked edge that is not allowed to be contracted.

2. In order to restrict the number of passengers in one vehicle, all adjacent edges of a node should be deleted when the capacity of the assigned vehicle is reached, see Fig. 4.10. If customers are allowed to fix each other as co-passenger, every contraction has to be checked, regarding the remaining capacity.

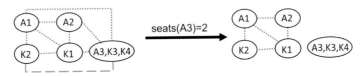

Figure 4.10: Graph transformation, when the number of maximal seats is reached.

In Fig. 4.11, the corresponding search tree of the constraint graph in Fig. 4.5 is shown. The children of a root can be obtained by contracting each green edge sequentially and coloring the contracted edge of the constraint graph in the root red, afterwards. An edge between the new node and another node is colored red in the

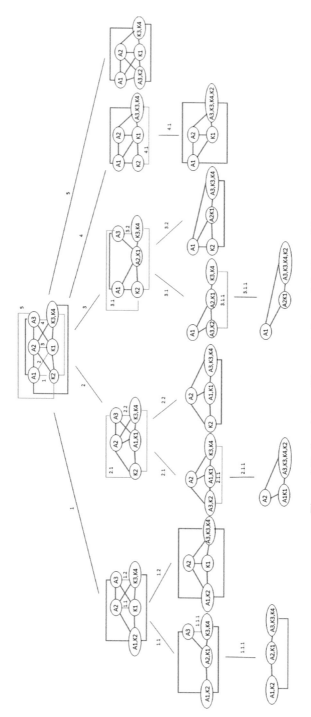

Figure 4.11: Search tree of the constraint graph in Figure 4.5.

constraint graph of a child, if at least one of the merged nodes had a red edge to it, and colored green, if both had green edges to it.

After the construction of the search tree is shown, the computation of routes and costs of each coalition are described. By assuming, that only shortest paths are driven between the starting and destination locations of each customer in the same coalition, the routes can be computed fast in case of small size groups of riders. A detailed description of the complexity can be found in [26].

One can even just enumerate all pickup and drop-off sequences, when the number of passengers in one coalition is small. Table 4.2 depicts the number of all possible sequences, with respect to the number of assigned passengers, to a common vehicle with five seats.

Table 4.2: Number of possibilities for feasible start and destination sequences with respect to the number of passengers

# passenger	1	2	3	4	5
# possibilities	1	6	90	2520	113.400

In order to find the best coalition structure and the corresponding best shortest path routes, a cost function has to be defined. As the cost of each feasible coalition structure is computed in the search process, this cost function can be individual for every coalition. Even an individual cost function for every customer can be defined. Bistaffa et al. suggest to use driving time, fuel cost and cognitive costs of the driver, to follow the route, as costs for the whole coalition. Branch and bound and parallel search on the constructed tree is used, to find the best coalition structure with minimal overall cost $CS^* = \text{argmin}_{CS \in FCS(G)} \sum_{S \in CS} c(S)$, where FCS is the set containing all feasible coalition structures CS generated by the search tree and $c(S)$ is equal to the cost of the coalition S. Further details can be found in [26].

4.5 Avoiding Congestions

Surely, there will be less accidents on streets, when all vehicles drive autonomously. Therefore, congestion caused by individual failure will be reduced. Another reason for congestion is the overuse of streets. Assuming that all vehicles are routed by one transportation system, there is a very simple solution for avoiding the overuse of streets and thereby congestion. The vehicles just have to be routed in a way, that the number of vehicles, using the same street at the same time, does not exceed the capacity of the street. This simple solution is also proposed in [33]. There, a capacity is assigned to each street, where the capacity is set to a congestion threshold, which can be determined by historic data. If this capacity is respected, the travel time of the vehicles, using this street, is set to the free-flow speed and infinity, else. Instead of just setting one threshold, the authors of [34] increase the traversal time of a street with the number of vehicles using it.

Other approaches include traffic predictions and re-routing the vehicles, that use predicted, congested streets. This attempt should be favoured when many road users have their own car and do not want to drive a proposed route. There are multiple references in literature, that concentrate just on traffic prediction, and some literature reference, including the predictions in the routing decision. As good traffic predictions are essential for re-routing strategies, various prediction concepts are described first.

The idea of using simulations to predict traffic is very promising at first glance. Unfortunately, many simulation environments like MAT.SIM base on too simple models and should therefore not be used for traffic predictions. Statistical methods like Auto-Regressive Integrated Moving Average (ARIMA) [35] and Kallmann Filter [36] are a better choice, but are not suited for big datasets. Further, they are very stationary and neglect dynamic local dependencies. Therefore, learning-based models, like support vector machine [37] and k-nearest-neighbor [38], are more accurate. Recently, deep learning models are most researched [39] and often have the best accuracy [40, 41, 42]. Long short-term memory (LSTM) models should be highlighted as solution concept at the moment. The work [43] is possibly well suited for traffic prediction in the era of autonomous vehicles, as it includes transportation requests. After having an overview over the traffic-prediction literature, rerouting concepts are described next. Just giving all vehicles an alternative route, when congestion on their route is predicted [44], is not a good idea, since new congestions can arise then. Therefore, vehicles using congested streets, should be re-routed in a balanced manner [45]. The works in [44] and [46] first predict the traffic on the streets, caused by vehicles, not participating in the transportation system, and then sequentially re-route vehicles, which use congested streets in the future. By sequential re-routing, the prediction can be updated before the next vehicle is re-routed and thereby new congestions are prevented. Pan et al. [46] introduced a fast and appropriate technique with multipath load balancing considering future vehicle positions. The combination of their technique and the accurate traffic prediction by LSTM was evaluated in [47].

If most people rather buy their own car instead of using the same transportation systems that shares them, privacy-retentive distributed algorithms [48] and cooperative re-routing strategies [49] should be studied.

4.6 A Rich Vehicle Routing Formulation

The heuristic in Section 4.4.1 already contains many requirements like heterogeneous vehicles, single or shared rides and co-passenger restrictions. However, it is not clear how re-optimization should take place in the online case. Further, the shortest path assumption prevents congestion avoidance. For this reason, the probably richest vehicle routing problem will be worked out in this section. First, a main demand-responsive vehicle routing problem based on multicommodity-coupled flows in time-expanded networks [50] is stated. After that, each requirement is formulated with constraints, which can be added to the main problem. At last, the overall formulation of the static problem is shown.

4.6.1 A Main Vehicle Routing Formulation

The main demand-responsive vehicle routing problem based on multicommodity-coupled flows in time-expanded networks is stated in this section. A Network $N = (V, A, c, b)$ consists of a directed Graph $G = (V, A)$ and a demand function $b : V \rightarrow \mathbb{Z}$ satisfying $\sum_{v \in V} b(v) = 0$. The demand $b(v)$ of a node $v \in V$ is called balance.

In multicommodity flow problems, an individual demand function $b_j : V \rightarrow \mathbb{Z}$ is assigned to each commodity j. A function $f : A \rightarrow \mathbb{R}_+$ that satisfies the balance constraints

$$\sum_{\{y:(x,y)\in A\}} f((x, y)) - \sum_{\{y:(y,x)\in A\}} f((y, x)) = b(x) \ \forall \ x \in V,$$

is called feasible network flow. The function value $f((x, y))$ of a feasible network flow is called flow on arc $(x, y) \in A$. A network flow that is feasible and minimizes the cost is called optimal. We write f^* for optimal network flows.

In multicommodity flow problems, a function $f_j : A \rightarrow \mathbb{R}_+$, which has to satisfy its individual balance constraints $b_j (x) \ \forall \ x \in V$ to be feasible, is assigned to each commodity j.

In contrast to most vehicle routing problems, the road network is not only represented by a directed graph $G = (V, A, c)$ with vertices $v \in V$, arcs $a \in A$ and arc costs $c(a) \in \mathbb{Z}_+$ for every arc $a \in A$. Travel times $\tau (a) \in \mathbb{Z}_+$ are added and explicitly captured in the graph structure.

This is done by expanding the graph in time $t = 0, \ldots, T$. In more detail, the corresponding time-expanded graph $G_T = (V_T, A_T, c_T)$ with time horizon T is obtained by setting:

$$V_T := \{x^t : x \in V, t = 0, \ldots, T \}$$
$$A_T := \{(x^t, y^{t+\tau((x,y))}) : (x, y) \in A, t = 0, \ldots, T - \tau^{((x, y))}\}$$
$$c_T(x^t, y^{t'}) := \begin{cases} 0, & \text{if } x = y \\ c(x, y), & \text{else.} \end{cases}$$

After introducing the underlying graph structure, transportation requests $r = (s, d, [p, q], \pi)$ are defined next. A transportation request contains a pickup location $s \in V$, a drop-off location $d \in V$, a time window $[p, q]$ and a profit $\pi \in \mathbb{Z}_+$.

The transportation requests r_i of customers $i \in I$ are served by vehicles $k \in K$, which are located in $s_k \in V$ at time 0 and have to end their tour in $d_k \in V$ at time T. Since a vehicle is only able to transport a bounded number of customers at one time, a limited capacity $cap \in \mathbb{Z}_+$, which is equal to the number of passenger seats, is assigned to each vehicle $k \in K$. In order to earn the profit π_i of a transportation request $r_i = (s_i, d_i, [p_i, q_i], \pi_i)$, a vehicle $k \in K$ has to transport customer i from s_i to d_i in the time window $[p_i, q_i]$.

The aim of the system is to route the vehicles, such that the profit of fulfilled request, minus the travelling costs of the vehicles, is maximized. As mentioned in the beginning of the section, this can be formulated as a network flow problem.

Given the underlying time-expanded graph $G_T = (V_T, A_T, c_T)$, transportation requests r_i of customers $i \in I$ and vehicles $k \in K$, the corresponding Network $\bar{N}_T = (\bar{V}_T \bar{A}_T, \bar{c}_T, (b_i)_{i \in I}, (b_k)_{k \in K})$ with customer demand functions b_i and vehicle demand functions b_k is constructed as follows:

First, a fictitious starting node $\bar{s}_i$ with balance 1 and fictitious end node $\bar{d}_i$ with balance −1 for every customer $i \in I$ is added to the time-expanded graph $G_T = (V_T, A_T, c_T)$. From the fictitious starting node $\bar{s}_i$ arcs to the pickup location $s^t_i, p_i \leq t \leq q_i$ with profit π_i and one arc to the fictitious end node $\bar{d}_i$ with profit 0 are added. This enables the system to deny or fulfill requests. The bound q_i of the time

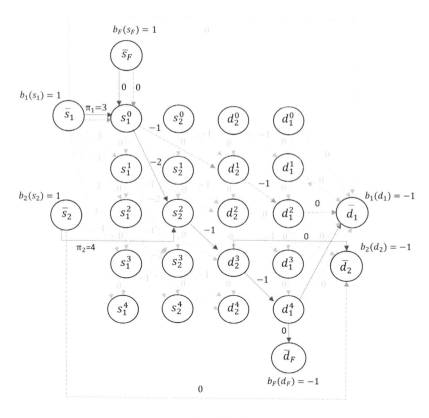

Figure 4.12: An extended network $\bar{N}_4 = (\bar{V}_4, \bar{A}_4, \bar{c}_4, b_1, b_2, b_F)$ with two customer requests $r_1 = (s_1, d_1, [0, 4], \pi_1 = 3)$, $r_2 = (s_2, d_2, [2, 3], \pi_2 = 4)$ and one vehicle F starting in s_1 and ending in d_1. In the dotted red solution customer 1 is picked up at time 0 and dropped off at time 4, while customer 2, is picked up at time 2 and dropped off at time 3.

window is met, by setting arcs form d_i^t with $p_i \le t \le q_i$ to $\bar{d}_i$ for every customer $i \in I$. For each vehicle, fictitious starting nodes $\bar{s}_k$ with balance 1 and fictitious end nodes $\bar{d}_k$ with balance -1 are created, too. By connecting $\bar{s}_k$ with s_k^0 and d_k^T with $\bar{d}_k$, each vehicle k starts at s_k at time 0 and ends at d_k at time T.

An example of a network and two solutions can be seen in Fig. 4.12. In the dotted red solution, customer 1 is picked up at time 0 and dropped of at time 4, while customer 2 is picked up at time 2 and dropped of at time 3.

In the following, the main vehicle routing problem, given the constructed network, is stated. Therefore, let the flow variables $f_i(a)$ on arcs $a \in \bar{A}_T$ correspond to request $r_i \in R$ and the flow variables $F_k(a)$ on arcs $a \in \bar{A}_T$ correspond to vehicle $k \in K$. Further, the out-going arcs of node $v^t \in \bar{V}_T$ are given by $\delta^-(v^t) = \{u^{t'} : (v^t, u^{t'}) \in A_T\}$ and the in-going arcs of v^t are given by $\delta^+(v^t) = \{u^{t'} : (u^{t'}, v^t) \in A_T\}$.

With this notation and the created network $\bar{N}_T$, the main demand-responsive routing problem can be solved by following integer program.

As requested, the objective function maximizes the profit of the fulfilled request, minus the travelling costs of the vehicles. In the dashed green solution of Figure 4.12, for example, the flow of request 1 uses the arc from the virtual starting node $\bar{s}_1$ to the initial position s_1^0 and thus yields a profit of $\pi_1 = 3$. Since the flow of request 2 uses the arc $(\bar{s}_2, \bar{d}_2)$, the request is denied and no profit is earned. In order to serve request 1, vehicle F has to use 4 arcs with overall cost of 2, which yields an overall profit of $1=3-2$. The constraints in (2) and (3) capture the starting and destination locations of the vehicles. In Fig. 4.12, the flow of vehicle F starts at $\bar{s}_k$ and ends at $\bar{d}_k$, since $\bar{s}_k$ has a balance of 1 and $\bar{d}_k$ has a balance of -1. To ensure connected vehicle routes, the flow conservation constraints in (4) are stated. For example, node d_2^1 in Figure 4.12 is entered and left by one vehicle flow, since the balance of node d_2^1 is equal to 0. Also, balance constraints (5) and (6) and flow conservation constraints (7) are needed for each customer. Constraint (8) ensures, that exactly one vehicle is assigned to each customer. Since every customer flow in the road network has to be carried by the assigned vehicle, constraint (9) couples these flows with the flow of the assigned vehicle, where the vehicle assignment of customer i is fixed by x_{ik} and the vehicle flow is only allowed to carry *cap* (capacity of vehicles) customers. In the dashed green solution of Fig. 4.12, for example, the flow of request 1 is coupled with the flow of vehicle F. At last, all variables have to be binary (10), (11) and (12).

4.6.2 Extensions of the Main Formulation

The main vehicle routing formulation is already demand responsive and respects time windows. In the next subsections, further desired requirements are formulated by constraints, which can be added to the main problem.

$$\max \sum_{i \in I} \sum_{a \in \delta^-(\bar{s}_i)} \pi_i \cdot f_i(a) - \sum_{k \in K} \sum_{a \in A} c(a) \cdot F_k(a) \tag{1}$$

$$\text{s.t.} \sum_{a\in\delta^-(\bar{s}_k)} F_k(a) = 1 \ \forall \ k \in K \tag{2}$$

$$\sum_{a\in\delta^+(\bar{d}_k)} F_k(a) = 1 \ \forall \ k \in K \tag{3}$$

$$\sum_{a\in\delta^-(v')} F_k(a) = \sum_{a\in\delta^+(v')} F_k(a) \forall v' \neq \bar{s}_k, \bar{d}_k \forall k \in K \tag{4}$$

$$\sum_{a\in\delta^-(\bar{s}_i)} f_i(a) = 1 \ \forall \ i \in I \tag{5}$$

$$\sum_{a\in\delta^+(\bar{d}_i)} f_i(a) = 1 \ \forall \ i \in I \tag{6}$$

$$\sum_{a\in\delta^-(v')} f_i(a) = \sum_{a\in\delta^+(v')} f_i(a) \forall v' \neq \bar{s}_i, \bar{d}_i, \forall i \in I \tag{7}$$

$$\sum_{k\in K} x_{ik} = 1 \ \forall \ i \in I \tag{8}$$

$$\sum_{i\in I} x_{ik} \cdot f_i(a) \leq cap \cdot F_k(a) \ \forall a \in A_T, \forall k \in K \tag{9}$$

$$x_{ik} \in \{0,1\} \forall i \in I, \forall k \in K \tag{10}$$

$$F_k(a) \in \{0,1\} \forall a \in \bar{A}_T, \forall k \in K \tag{11}$$

$$f_i(a) \in \{0,1\} \forall a \in \bar{A}_T, \forall i \in I \tag{12}$$

4.6.2.1 Adding Heterogeneous Vehicles

By giving the vehicles $k \in K$ different individual capacities cap_k, a first step towards heterogeneity can be done. Further, the vehicle choice of a customer i can be integrated by setting the variables $x_{ik} = 0$ for every vehicle k, that does not satisfy the requested properties. If, for example, a customer i does not like to drive with vehicle k, just add the constraint $x_{ik} = 0$.

4.6.2.2 Booking a Ride for a Whole Group

If a customer books a ride for a whole group of people with same pickup and destination location, more capacity is occupied by the corresponding flow. Therefore, the corresponding flow has to be multiplied by the number of group members in the couple constraint. Let g_i be the number of group members of request i, then the following couple constraints can consider more than one customer per request.

$$\sum_{i\in I} x_{ik} \cdot f_i(a) \cdot g_i \leq cap \cdot F_k(a) \forall a \in A_T, \forall k \in K.$$

4.6.2.3 Giving the Opportunity to Choose Single or Shared Modes

In order to guarantee single-mode rides, the capacity of the assigned vehicle has to be completely occupied by single-mode passengers. This can be done by adding a variable *single$_i$* for every customer i, which is set to one, if customer i wants a single-mode ride and zero, else. Further, the coupling constraints have to be changed to

$$\sum_{i\in I} x_{ik} \cdot f_i(a) \cdot single_i \cdot cap + \sum_{i\in I} x_{ik} \cdot f_i(a) \cdot (1 - single_i) \le cap \cdot F_k(a),$$

$$\forall \, a \in A_T, \, \forall \, k \in K.$$

4.6.2.4 Congestion Avoidance

The idea of Rossi et al. to avoid congestion [33] can be simply added to the main problem by giving every arc $a \in A_T$ a congestion threshold $ct(a)$ and adding following constraint:

$$\sum_{k\in K} F_k(a) \le ct(a) \forall a \in A_T.$$

4.6.2.5 Co-passenger Restrictions

Co-passenger restrictions can be accomplished by introducing fixed variables p_{ij} for each pair of customers i and j, where $p_{ij} = 1$, when one of the two customers does not want to drive with the other, and zero, else. Further, the following constraint for each customer pair has to be added:

$$\sum_{k\in K} x_{ik} \cdot x_{jk} \cdot p_{ij} \le 0 \, \forall \, i \ne j \in I$$

The following IP optimizes the static problem with all requirements.

4.6.2.6 Dynamic Extension

In contrast to the static case, where all request are known before, the requests in the online case become known over time. More accurate, every request has a release date $t_i^r \in \{1, \ldots, T\}$. Like in [50], the decision

$$\max \sum_{i\in I} \sum_{a\in\delta^-(\bar{s}_i)} \pi_i \cdot f_i(a) - \sum_{k\in K} \sum_{a\in A} c(a) \cdot F_k(a) \tag{1}$$

$$\text{s.t.} \sum_{a\in\delta^-(\bar{s}_k)} F_k(a) = 1 \, \forall \, k \in K \tag{2}$$

$$\sum_{a\in\delta^+(\bar{d}_k)} F_k(a) = 1 \, \forall \, k \in K \tag{3}$$

$$\sum_{a\in\delta^-(v')} F_k(a) = \sum_{a\in\delta^+(v')} F_k(a) \, \forall \, v' \ne \bar{s}_k, \bar{d}_k, \, \forall \, k \in K \tag{4}$$

$$\sum_{a\in\delta^-(\bar{s}_i)} f_i(a) = 1 \; \forall \, i \in I \tag{5}$$

$$\sum_{a\in\delta^+(\bar{d}_i)} f_i(a) = 1 \; \forall \, i \in I \tag{6}$$

$$\sum_{a\in\delta^-(v')} f_i(a) = \sum_{a\in\delta^+(v')} f_i(a) \; \forall \, v' \neq \bar{s}_i, \bar{d}_i \; \forall \, i \in I \tag{7}$$

$$\sum_{k\in K} x_{ik} = 1 \; \forall \, i \in I \tag{8}$$

$$\sum_{i\in I} x_{ik} \cdot f_i(a) \cdot single_i \cdot \max\{cap_k, g_i\}$$
$$+ \sum_{i\in I} x_{ik} \cdot f_i(a) \cdot (1-single_i) \cdot g_i \tag{9}$$
$$\leq cap_k \cdot F_k(a) \; \forall a \in A_T, \forall k \in K$$

$$\sum_{k\in K} x_{ik} \cdot x_{jk} \cdot p_{ij} \leq 0 \, \forall i \neq j \in I \tag{10}$$

$$\sum_{k\in K} F_k(a) \leq ct(a) \, \forall a \in A_T \tag{11}$$

$$x_{ik} \in \{0,1\} \, \forall i \in I, \forall k \in K \tag{12}$$

$$F_k(a) \in \{0, 1\} \, \forall a \in \bar{A}_T, \forall k \in K \tag{13}$$

$$f_i(a) \in \{0, 1\} \, \forall a \in \bar{A}_T, \forall i \in I \tag{14}$$

of accepting or denying an incoming request can be done by solving static integer problems every time step new requests arrive. The IP of the static case just has to be adjusted to the current position of the customer and vehicles. An example can be found in [50].

4.7 Distributed Vehicle Routing Solutions

Unfortunately, at present, the IP stated in Section 4.6 cannot be solved for many requests in reasonable time. Distributed algorithms are very promising, since they scale well with the number of request. Possibly, the numerous constraints can be exploited and used to distribute the computation. This section, therefore gives an overview of distributed algorithms.

4.7.1 Ranking

A very simple distributed algorithm, based on rankings, is introduced in [51]. There, each customer request is sent to all, or rather a subset of vehicles. The vehicles then compute their best, feasible service sequence, given the already-assigned requests and the new one, where a feasible service sequence satisfies service degradation constraints and the capacity constraint of the vehicle. Lastly, all vehicles sent their solution, with corresponding costs, to the customer application, which chooses the vehicle with cheapest costs. A similar method can be found in [52].

The work of [53] extends this idea by introducing fog nodes. Instead of sending the request directly to vehicles, they are first sent to the best k fog nodes, which represent a subset of vehicles near the fog node. The best k fog nodes are determined by computing the load of service, which considers the detour ratio and the average number of on-board riders. Each fog node sends the best vehicle, with cheapest detour cost, to the cloud, which makes decision about the assignment.

Instead of fog nodes, a broker is used in [54], to determine the subset of vehicles that fulfil the hard vehicle preferences of the customer. All vehicles, fulfilling the constraints, compute their utility of inserting the new request and sent it to a planner. This planner selects the most appropriate offers and forwards them to the client, which may choose his favourable vehicle. In the filtering, not only the customer preferences, but also aims of the system, like the minimization of the number of used vehicles, are considered. Further examples, that give the customer the ability to choose from offers, can be found in [55] and [56].

Ranking methods can fulfil many constraints, like time windows, co-passenger restrictions, individual rides and vehicle choices, since each vehicle can check these constraints, when the request is added in the schedule. However, each request can only be processed sequentially.

4.7.2 Reinforcement Learning

Recently, reinforcement learning is used to route vehicles. At first glance, this seems very promising, since future travel demands are included in the models. However, it is very hard or not even appropriate, to adapt these algorithms to more constraint vehicle routing problems. For example, in [57], only traveling salesman problems and delivery problems are solved.

In [58], the transportation of customers is considered. However, time windows are neglected, heterogeneous vehicles are not concerned, zone-to-zone transportation is assumed and only the transportation of at most two customers, by one vehicle, is allowed.

The scenario in [59] is more complex, since the vehicles can have different capacities and transport at most this number of customers. Further, it is more customer-friendly, since the waiting time of customers and the additional time by ride sharing is included in the reward function of the vehicles.

To our knowledge, there are no routing solutions, based on reinforcement learning, considering hard constraints. Since the development of an appropriate masking scheme is very difficult [57], we are unable to evaluate the appropriateness of reinforcement learning to our rich problem at this moment.

4.7.3 Game Theory

Negotiation should not be neglected in the decentral routing of vehicles. The authors of [58] recognized this and propose a potential game and binary log-linear learning, to prevent agents to adapt to the same request.

Another game-theoretic approach for negotiation are auctions. In [51], the vehicles bid on the request sequentially, where the bids are computed by optimization strategies or heuristics.

In the case of unlimited vehicle capacity, a genetic algorithm is constructed to compute the bids [60]. Further, the authors also use auctions to redistribute requests. However, these kind of auctions do not take the advantage of servicing a combination of transportation requests into account. Therefore, the authors of [61] propose a combinatorial auction, where each vehicle bids on each of the 2^{n-1} possible subsets, where n is the number of requests. This idea however, is not applicable for a high amount of requests. That is why many authors in the field of collaborating vehicle routing shrink the number of offered subsets. The constraints of our rich IP already shrinks the set of feasible subsets and makes the idea of combinatorial auctions more welcome.

Further shrinkage can be achieved by letting the vehicles propose subsets [62] or by constructing only attractive subsets in a central manner [63]. For example, Ganstarer et al. [63] use genetic algorithms to construct attractive bundles, without the knowledge of already-assigned requests.

Their idea can be adapted to our rich vehicle routing problem by setting the fitness function value to zero, if constraints are violated. Since the crossovers and the mutations mainly consider geographical information, they possibly have to be extended.

4.7.4 Cluster First, Route Second

The idea of first building clusters of requests and then route the vehicles according to them, is followed by [64]. In order to optimize the delivery of goods, first, a clustering problem is solved. After that, the routes of each cluster are optimized.

A distributed heuristic, based on randomization and local optimization, can be used to solve the central clustering problem. In each iteration, a local problem considering only a random subset of vehicles with actual assigned customers is solved. When each vehicle was involved in a local optimization iteration at least once and the corresponding central clustering IP is feasible, their procedure outputs a feasible solution. After the clustering phase, each vehicle optimizes the route for one cluster. In order to apply the ideas of [64] to our rich vehicle routing problem, first, a central clustering problem, that considers the various hard constraints, has to be developed. This problem can be solved analogously by randomization and local optimization. At last, the routes can be optimized considering traffic.

4.8 Conclusion

In this chapter, a desired autonomous transportation system is introduced and mathematically modelled by an IP-formulation. First of all, it should be verified by a user survey, whether the collected requirements are appropriate. After that, the requirements should possibly be adjusted.

In order to solve the resulting problem in reasonable time, heuristics have to be invented. Distributed algorithms are very promising, since they scale well with the number of requests and vehicles. The review of distributed algorithms can give readers first ideas for the development of algorithms, solving such a rich vehicle routing problem. The combination of machine learning and auctions, in particular the computation of bids by machine learning, is to be emphasized.

Giving a customer many alternatives to choose from, can frustrate him, since he does not know which resources are occupied and which not. It might happen, that the user has to send many request till he finds an available transport.

Therefore, the implementation of a flexible mobility-on-demand system [65], which gives the customer appropriate transport proposals with fair cost allocations, is the goal of the chapter authors and should be studied in more detail. In literature, this problem is very abstracted. For example, to the best of our knowledge, all authors assume an immediate answer from the customer. In contrast, we generated in [66] the assortments of travel offers considering the choices of all customers simultaneously, which makes the application more practical, as users need time to choose a travel option.

Acknowledgment

This project was funded by the Federal Ministry of Transport and Digital Infrastructure (BMVI) with the funding guideline "Automated and Connected Driving" under the funding code 16AVF2134C.

References

1. Pavone, M. 2015. Autonomous mobility-on-demand systems for future urban mobility. *In*: Maurer, M., Gerdes, J., Lenz, B., Winner, H. (Eds.). Autonomous Driving. Springer, Berlin, Heidelberg.
2. Fagnant, D.J. and K.M. Kockelman. 2018. Dynamic ride-sharing and fleet sizing for a system of shared autonomous vehicles in Austin, Texas. *Transportation*, 45, 143–158.
3. Uber. 2019, September 13. Retrieved from https://www.uber.com/fr/fr.
4. Lyft. 2019, September 13. Retrieved from https://www.lyft.com/rider.
5. DiDi. 2019, February 25. Retrieved from https://www.didiglobal.com.
6. Careem. 2019, February 25. Retrieved from https://www.careem.com/
7. MOIA. 2019, September 13. Retrieved from https://www.moia.io/de-DE.
8. Perera, S., C. Ho and D. Hensher. 2019. Resurgence of Demand Responsive Transit Services. Insights from BRIDJ Trials in Inner West of Sydney, Australia.
9. Monheim. 2020, February 25. Retrieved from https://www.electrive.com/2019/03/28/five-fully-autonomousbuses-to-operate-in-monheim-germany/
10. Enoch, M.P., S. Potter, G. Parkhurst and M. Smith. 2004. INTERMODE: Innovations

in demand responsive transport. Report for Department for Transport and Greater Manchester Passenger Transport Executive.

11. Howard, D. and D. Danielle. 2014. Public perceptions of self-driving cars: The case of Berkeley, California. 93th Annual Meeting of the Transportation Research Board.

12. Gargiulo, E., R. Giannantonio, E. Guercio, C. Borean and G. Zenezini. 2015. Dynamic ride sharing service: Are users ready to adopt it? *Procedia Manufacturing*, 3, 777–784.

13. Wang, Y., R. Kutadinata and S. Winter. 2019. The evolutionary interaction between taxi-sharing behaviours and social networks. *Transportation Research Part: A Policy and Practice*. 119, 170–180.

14. Amey, A.P. and R.G. Mishalani. 2010. Real-time ridesharing—The opportunities and challenges of utilizing mobile phone technology to improve rideshare services. *Transportation Research Record*. 2217, 103110.

15. Schoettle, B. and M. Sivak. 2014. A survey of public opinion about autonomous and self-driving vehicles in the U.S., the U.K., and Australia. Retrieved from https://deepblue.lib.umich.edu/bitstream/handle/2027.42/108384/103024.pdf.

16. Zmud, J., I. Sener and J. Wagner. 2016. Consumer acceptance and travel behavior impacts of automated vehicles. Retrieved from https://d2dtl5nnlpfr0r.cloudfront.net/tti.tamu.edu/documents/ PRC-15-49-F.pdf.

17. Merat, N., R. Madigan and S. Nordhoff. 2017. Human factors, user requirements, and user acceptance of ride-sharing in automated vehicles. International Transport Forum Discussion Paper 10.

18. Piao, J., M. McDonald, N. Hounsell, M. Graindorge, T.,Graindorge and N. Malhene. 2016. Public views towards implementation of automated vehicles in urban areas. *Transportation Research Procedia*, 14, 2168–2177.

19. Nordhoff, S., B. van Arem and R. Happee. 2016. A conceptual model to explain, predict, and improve user acceptance of driverless 4P vehicles. *Transportation Research Record*, 2602: 6067.

20. Lavieri, P. and C. Bhat. 2019. Modeling individuals willingness to share trips with strangers in an autonomous vehicle future. *Transportation Research Part: A Policy and Practice*, 124, 242–261.

21. Pfleging, B., M. Rang and N. Broy. 2016. Investigating user needs for non-driving-related activities during automated driving. Proceedings of the 15th International Conference on Mobile and Ubiquitous Multimedia (MUM '16). 91–99.

22. Gossen, M. and Wirtschaftsforschung Berlin. 2012. Nutzen statt be - sitzen - Motive und Potenziale der internetgesttzten gemeinsamen Nutzung am Beispiel des Peer-to-Peer Car-Sharing. Schriftenreihe des IW, 202, 12.

23. Cyganski, R. 2016. Automated Vehicles and Automated Driving from a Demand Modeling Perspective. *In*: Maurer, M., Gerdes, J., Lenz, B., Winner, H. (Eds.). Autonomous Driving. Springer, Berlin, Heidelberg.

24. Hyland, M.F. and H.S. Mahmassani. 2017. Taxonomy of shared autonomous vehicle fleet management problems to inform future transportation mobility. *Transportation Research Record*, 2653(1), 26–34.

25. Parragh, S. 2011. Introducing heterogeneous users and vehicles into models and algorithms for the dial-a-ride problem. *Transportation Research Part C-emerging Technologies*, 19, 912–930.

26. Bistaffa, F., A. Farinelli, G. Chalkiadakis and S. Ramchurn. 2017. A cooperative game-

theoretic approach to the social ridesharing problem. *Artificial Intelligence*, 246, 86–117.

27. Levinger, C., A. Azaria and N. Hazon. 2018. Human satisfaction as the ultimate goal in ridesharing. ArXiv, abs/1807.00376.

28. Hendawi, A.M., A. Rustum, A.A. Ahmadain, D. Hazel, A. Teredesai, D. Oliver, M.H. Ali and J.A. Stankovic. 2017. Smart personalized routing for smart cities. IEEE 33rd International Conference on Data Engineering (ICDE). 1295–1306.

29. Saisubramanian, S., C. Basich, S. Zilberstein and C. Goldman. 2019. The value of incorporating social preferences in dynamic ridesharing. ICAPS 2019 Workshop SPARK Blind Submission.

30. Aiko, S., P. Thaithatkul and Y. Asakura. 2018. Incorporating user preference into optimal vehicle routing problem of integrated sharing transport system. *Asian Transport Studies*, 5, 98–116.

31. Wang, Y., R. Kutadinata and S. Winter. 2019. The evolutionary interaction between taxi-sharing behaviours and social networks. *Transportation Research Part A: Policy and Practice*, 119, 170–180.

32. Middleton, S. and J. Zhao. 2019. Discriminatory attitudes between ride sharing passengers. *Transportation*, vol n pp

33. Rossi, F., R. Zhang, Y. Hindy and M. Pavone. 2018. Routing autonomous vehicles in congested transportation networks: Structural properties and coordination algorithms. *Autonomous Robots*, 42(7), 1427–1442.

34. Salazar, M., M. Tsao, I.P. Aguiar, M. Schiffer and M. Pavone. 2019. A congestion-aware routing scheme for autonomous mobilityon-demand systems. 18th European Control Conference (ECC). 3040–3046.

35. Mai, T., B. Ghosh and S. Wilson. 2014. Short-term traffic-flow forecasting with auto-regressive moving average models. Proceedings of the ICE – Transport, 167, 232–239.

36. Shekhar, S. and B. Williams. 2008. Adaptive seasonal time series models for forecasting short-term traffic flow. *Transportation Research Record*, 2024, 116–125.

37. Castro-Neto, M., Y.-S. Jeong, M.-K. Jeong and L. Han. 2009. Online-SVR for short-term traffic flow prediction under typical and atypical traffic conditions. *Expert Systems with Applications*, 36, 6164–6173.

38. Luo, X., D. Li, Y. Yang and S. Zhang. 2019. Spatiotemporal traffic flow prediction with KNN and LSTM. *Journal of Advanced Transportation*, 1-10.

39. Do, L., N. Taherifar and H. Vu. 2018. Survey of neural network-based models for short-term traffic state prediction. *Wiley Interdisciplinary Reviews: Data Mining and Knowledge Discovery*, 9.

40. Isrehin, N.O., A.F. Klaib and A.A. Magableh. 2019. Intelligent transportation and control systems using data mining and machine learning techniques: A comprehensive study. *IEEE Access*, 7, 49830–49857.

41. Yao, H., X. Tang, H. Wei, G. Zheng and Z. Li. 2018. Revisiting spatial-temporal similarity: A deep learning framework for traffic prediction. Proceedings of the AAAI Conference on Artificial Intelligence, 33, 5668–5675.

42. Lv, Y., Y. Duan, W. Kang, Z. Li and F. Wang. 2015. Traffic flow prediction with big data: A deep learning approach. *IEEE Transactions on Intelligent Transportation Systems*, 16, 865–873.

43. Liao, B., J. Zhang, C.H. Wu, D. McIlwraith, T. Chen, S. Yang, Y. Guo and F. Wu. 2018. Deep sequence learning with auxiliary information for traffic prediction. KDD. Vol n pp

44. Wilkie, D., J.P. Berg, M.C. Lin and D. Manocha. 2011. Selfaware traffic route planning. Proceedings of the Twenty-Fifth AAAI Conference on Artificial Intelligence. 2.
45. Souza, A., R.Yokoyama, G. Maia, A. Loureiro and L. Villas. 2016. Real-time path planning to prevent traffic jam through an intelligent transportation system. *IEEE Symposium on Computers and Communication (ISCC)*, 726–731.
46. Pan, S.J., M.A. Khan, I.S. Popa, K. Zeitouni and C. Borcea. 2012. Proactive vehicle re-routing strategies for congestion avoidance. IEEE 8th International Conference on Distributed Computing in Sensor Systems, 265–272.
47. Perez-Murueta, P., A. Gmez-Espinosa, C. Crdenas and M. Gonzlez-Mendoza. 2019. Deep learning system for vehicular re-routing and congestion avoidance. *Applied Sciences*, 9, 2717.
48. Pan, S.J., I.S. Popa and C. Borcea. 2017. DIVERT: A distributed vehicular traffic re-routing system for congestion avoidance. *IEEE Transactions on Mobile Computing*, 16, 58–72.
49. Yan, L., W. Hu and S. Hu. 2018. SALA: A self-adaptive learning algorithm towards efficient dynamic route guidance in urban traffic networks. *Neural Processing Letters*, 1–25.
50. Bsaybes, S., A. Quilliot and A. Wagler. 2018. Fleet management for autonomous vehicles using multicommodity coupled flows in time-expanded network. 17th International Symposium on Experimental Algorithms, SEA 2018. 103.
51. d'Orey, P.M., R. Fernandes and M. Ferreira. 2012. Empirical evaluation of a dynamic and distributed taxi-sharing system. 15th International IEEE Conference on Intelligent Transportation Systems. 140–146.
52. Bathla, K., V. Raychoudhury, D. Saxena and A.D. Kshemkalyani. 2018. Real-time distributed taxi ride sharing. 21st International Conference on Intelligent Transportation Systems (ITSC). 2044–2051.
53. Lai, Y., F. Yang, L. Zhang and Z. Lin. 2018. Distributed public vehicle system based on fog nodes and vehicular sensing. *IEEE Access*, 6, 22011–22024.
54. Cubillos, C., F. Guidi-Polanco and C. Demartini. 2004. Multiagent infrastructure for distributed planning of demand-responsive passenger transportation service. *IEEE International Conference on Systems: Man and Cybernetics*, 2, 2013–2017.
55. Winter, S. 2008. Intelligent self-organizing transport. KI, 22, 25–28.
56. Winter, S. and S. Nittel. 2006. Ad hoc shared-ride trip planning by mobile geosensor networks. *International Journal of Geographical Information Science*, 20, 899–916.
57. Nazari, M., A. Oroojlooy, L.V. Snyder and M. Takc. 2018. Deep reinforcement learning for solving the vehicle routing problem. ArXiv, abs/1802.04240.
58. Rahili, S., B. Riviere, S. Olivier and S. Chung. 2018. Optimal routing for autonomous taxis using distributed reinforcement learning. IEEE International Conference on Data Mining Workshops (ICDMW). 556–563.
59. Al-Abbasi, A.O., A. Ghosh and V. Aggarwal. 2019. Deep Pool: Distributed model-free algorithm for ride-sharing using deep reinforcement learning. ArXiv, abs/1903.03882.
60. Lon, R.R., J. Branke and T. Holvoet. 2017. Optimizing agents with genetic programming: An evaluation of hyper-heuristics in dynamic real-time logistics. *Genetic Programming and Evolvable Machines*, 19, 93–120.
61. Gomber, P., C. Schmidt and C. Weinhardt. 1997. Elektronische Mrkte fr die dezentrale Transportplanung. *Wirtschaftsinformatik*, 39, 137–145.
62. Dijkman-Krajewska, M. and H. Kopfer. 2006. Collaborating freight forwarding enterprises: Request allocation and profit sharing. *OR Spectrum*, 28, 308–317.

63. Gansterer, M. and R.F. Hartl. 2018. Centralized bundle generation in auction-based collaborative transportation. *OR Spectrum*, 40, 613–635.
64. Abbatecola, L., M.P. Fanti, G. Pedroncelli and W. Ukovich. 2018. A new cluster-based approach for the vehicle touting problem with time windows. IEEE 14th International Conference on Automation Science and Engineering (CASE). 744–749.
65. Atasoy, B., T. Ikeda and M.E. Ben-Akiva. 2016. Optimizing a flexible mobility on demand system. *Transportation Research Record: Journal of the Transportation Research Board*, 2536, 76–85.
66. Dziubany, M., J. Schneider, A. Schmeink and G. Dartmann. 2020. Optimization of a CPSS-based Flexible Transportation System. 9th Mediterranean Conference on Embedded Computing (MECO). 1–5.

Future Urban Mobility: Designing New Mobility Technologies in Open Innovation Networks

David Hedderich*, Markus Kowalski and Volker Lücken

e.GO Mobile AG, Campus-Boulevard 30, 52072 Aachen, Germany

1.1 Introduction

This chapter examines how different organizations can cope with and handle the temporal mode of the future in urban mobility in the light of the digital transformation [1, 2]. In terms of the importance of organizing future mobility in ecosystems, we still lack understanding of which structures, processes and culture we need in mobility networks to create technologies and innovations. One phenomenon that seems pervasive of today's organizational landscape in the mobility ecosystem is the challenge to cope with issues imposed by an ever-increasing need to face customized services and digitalization [3, 4]. Hence, to address challenges as significant as high levels of emission and increasing traffic congestion of today's upscaling cities we need to work on new technologies and transorganizational forms of cooperation to get an actively managed urban area. Hence, to keep this urban mobility performance indicator in mind, public transportation (PT) should remain the focus within the vehicle-based modalities of urban mobility planning [5]. Apparently, this imagined state will not change with the possible trajectory that the class of new and smaller robo-taxis, which may cannibalize urban transport in future cities by mimicking traffic patterns of individual mobility. Instead of removing public transport and common mobility solutions, new smart technologies must be reinforced – such as in automated driving, connectivity and digitalization – to fulfil the needs of the customer and solve challenges in urban mobility. In our study we address these topics through an explorative case study [6], observing how an actively orchestrated innovation

---

*Corresponding author: david.hedderich@rwth-aachen.de

network develops new urban management concepts to generate an intelligent and sustainable city of the future by means of venturing towards becoming a digitally transformed city (e.g. automated cars and on-demand-mobility). Our analysis shows, that innovation networks offer an at least partially generalizable account of how to engage different actors who cooperate in terms of finding solutions for challenges in future urban mobility [3, 7, 8]. Building on these observations, we add to the discussion in existing literature by (1) providing a framework which utilizes automated and electrified on-demand buses to pool together several passengers for a certain direction of transport managing future urban mobility and by (2) revealing structures and processes of how to manage the digital transformation and to provide a platform for emerging technologies, such as artificial intelligence (AI). Moreover, (3) our insights at least indirectly point attention to cities as informative urban laboratories to engage with the digitalization, offering an alternative setting to contemplate when compared to more common settings of digitalization efforts (e.g. such as companies or research institutes).

5.2 Contemporary Challenges and Concepts in Urban Mobility

The transformation of urban mobility is present – it is important to meet the requirements of citizens and organizations and to work together on solutions as part of realistic target scenarios. This also inheres an inclusion of the real-world complexity and interaction within the city as a system. We first motivate different relevant challenges and trends in urban mobility, which will be afterwards discussed by insights of current literature on urban ecosystems.

5.2.1 Urbanization as a Driver for Changing the Ecosystem of Mobility

The worldwide ongoing trend of urbanization imposes challenges on several layers of the mobility system, e.g. facing the perspective that about 85% of urban population will live in urban areas by 2050 [9]. With the given limitations of urban space and area, the density of both city inhabitants and traffic will increase. This affects several aspects of citizens' daily life and mobility (e.g. corresponding travel times or the convenience level). Further, high traffic densities and the required infrastructure exhibit a strong impact on shape and structure of the general cityscape. For example, significant traffic volumes require the allocation of traffic space and capacities, e.g. by increasing the number of available lanes for vehicles; however, this does not represent a viable option in the present city environment. Besides, for example, a large number of vehicles which are parking in the street have a negative effect on living space in many cities, especially historic cities in Europe that only exhibit limited space in-between the areas occupied by buildings are most of all affected by these actions. Finally, emissions of nitrogen oxides and fine particular matter also affect the real quality of life in cities in a negative way.

5.2.2 Intermodal Transition

Hence, to address the challenges, we require a fundamental change of old paradigm in the field of mobility. Intermodal transport must be implemented in an urban system and beyond. Furthermore, we need digital components which facilitate this change in paradigm from travelling alone from point A to B towards using intermodal solutions in a community manner. While intermodality is proclaimed as an essential component of efficient mobility systems, the level of implementation in practice is very low, as shown by several studies. For example, [10] reports a very low general level of intermodality in urban traffic of only 2% (referring to two or more modalities within a specific trip). For German cities with a population of more than 500,000 people, studies estimate the intermodality share between 5.8% [10] and 6.6% [11] based on mobility data, while for smaller cities with 100,000 to 500,000 inhabitants, [11] it reports a level of only 2.8%. Hence, reliable intermodality solutions are required to tackle the significant private use of motorized individual vehicles (MIV), which represent a substantial load for the urban traffic system [11].

Seamless intermodal transportation can be facilitated both by a transformation of urban infrastructure and a digitalization to interact with the mobility and traffic system. One solution, often envisioned as a central future mobility component, are urban mobility hubs. These hubs can constitute an infrastructural basis of intermodality, by combining different modalities (such as PT with rail and bus transportation [12], MIV and new mobility services) together with additional services provided at the point of transition (such as shopping, office and housing units [12]). These hubs can be placed at the city entrance, or also in central or suburban areas [12]. For MIV, these hubs can also be extended by implementing

Figure 5.1: Vision of mobility hubs as infrastructure for urban mode transition.

parking capabilities, constituting a dimension of transition similar to classical park-and-ride (P&R) approaches. By using this concept, the potential of P&R can be leveraged by minimizing inner city vehicle density, as shown in literature [13, 14].

By analyzing mobility systems (also refer to Section 5.2.4), Parkhurst and Meek state that "(t)ravellers are more likely to interchange early in the journey if the public transport mode offers an attractive journey time or reliability advantages (…)" [14]. It means, that shaping the future role of MIV in cities can have an impact on a seamless and comprehensive multimodal mobility system which can be placed at the forefront of the transformation process. At the same time, the efficiency of the mode transitions has to be ensured [15] in addition to the efficiency of the modes itself. Finally, by providing a transition of the majority of inbound traffic to emission-free inner-city modes and fleets, the scaling effects can be leveraged to accelerate the reduction of both local emissions and the global carbon dioxide balance.

5.2.3 Traffic Congestion and Flow Optimization

The alleviation of traffic load in inner cities by an efficient aggregation, bundling and transition of mobility has been motivated before. At the same time, the aim to efficiently exploit existing traffic infrastructure is still highly relevant in this transformation process and change of paradigm. On the one hand it includes topics of traffic flow optimization, to increase the robustness of the traffic system and decrease its sensitivity to disturbances. On the other hand, the general routing of vehicles can be optimized by using data-based approaches combined with knowledge of spatial and temporal patterns of urban traffic. Given the vision of a Smart City, the data can be gathered by infrastructural measures that facilitate the connectivity of infrastructure together with different sensing components for a real-time traffic sensing (with a further overview provided in [16]). Using crowd-sourced traffic measurements and historical data, a comprehensive traffic status of the urban environment is gathered and it can be combined with further forecast, prediction or AI methods for the traffic status (e.g. [17, 18]). The basis for a smart traffic routing can be built upon a city-wide optimization (e.g. [19, 20]) based on traffic patterns and events.

5.2.4 Metrics for Novel Mobility Systems and Services

While new approaches arise in the field of urban mobility, driven by specific technology components such as electrical powertrains and automated driving functionalities, the prevalent mobility system also needs to optimize the service level to exploit its potentials and provide an accessibility in the society [19]. Bertolini and le Clercq define the accessibility as "(…) the amount and the diversity of 'spatial opportunities' that can be reached within a certain amount of time" [21]. This corresponds to the ultimate goal of mobility to pursue "spatially disjointed activities" [21].

In general, mobility services have the target to minimize travel times, however, the optimization cost function for the overall system should also incorporate further metrics, to comprehend: (1) sustainability metrics (e.g. gasoline consumption, CO_2 emissions [21, 22, 23]); (2) a minimization of overall (individual) trip frequencies [19, 21]; (3) a reduction of traffic congestion and its economic and social costs [23] (which can incorporates the relation between individual mobility and PT [22]); and (4) the previously mentioned aspect of general accessibility [19]. Urban mobility systems often exhibit intermodal concepts in their solution space [19, 24]. The different modes can comprise new customized mobility services, driven by digitalization and a dynamic on-demand and possible multi-passenger scheduling, as presented with demand-responsive transport (DRT) in the following chapter. Their integration constitutes an important building block for the desired effective and reliable (and therefore termed as seamless) mobility system.

5.2.5 Automated Mobility

The dimension of automated driving, which we target for SAE level 4 (highly automated driving) [25] capabilities in this chapter, also holds implications for the mobility system. While the basic differentiation whether a vehicle still possesses a human driver or not is only marginally relevant for the mobility system. New capabilities in terms of a larger-area and around-the-clock coverage of public transportation arise, together with the possibility to scale new (e.g. smaller) vehicle concepts without a driver, which we discuss later in this chapter. In the case of robo-taxis without passenger pooling, basically the same principles of traffic patterns apply as with MIV (if no further optimizations and paradigmatic changes are applied). Thus, their potential for change as part of the mobility system is initially limited. Only in terms of alleviation of parked cars in the inner city (either in a tradeoff for parking traffic or in a sharing concept), the robo-taxis can provide a significant impact on urban mobility. Therefore, automated mobility has to be combined with tailored, flexible and intelligent services that are capable to exploit the different characteristics of these new types of vehicles and integrate them into comprehensive and cross-technological systems.

5.3 Mobility-as-a-Service with Automated Electric Buses

Towards this end, there exist multiple challenges for urban mobility in the future that base on the failure of the classic mobility paradigm: Privately-owned MIV, which are supplemented by public mass transport vehicles like trains, subways or busses. As urbanization is accompanied by an ever-increasing metric for the classic mobility paradigm of vehicles per 1,000 inhabitants in OECD and non-OECD countries, new solutions are necessary to manage future mobility in urban areas [26]. Without a paradigm shift towards a more sustainable mobility system, cities all over the world will not have a worthwhile future in terms of a livable environment. In fact, such a possible paradigm shift is often anticipated to be established by the technology of automated and connected driving as well as digitalization. For automated mobility, a differentiated view on this potential has also been provided in Section 5.2.5. Given these prospects, these technological parts are named as a 'future solution' in current research literature as well as in popular science. Precisely, automated and connected vehicles are perceived to be utilized in a digital DRT system that enables mobility at the fingertips. The accompanying research on mobility ecosystems focuses still on the examples of Uber and Lyft which are inspired by a ride-sharing approach of driverless robo-taxi-fleets. In the end, both organizations bring their passengers from A to B like a classic taxi service – and have been offering this service for a couple of years now [27]. Thus, the ride-sharing concept is adapted to the ride-pooling concept, which bundles multiple users that have a similar demand of a specific transportation route. Beyond that, they do not have necessarily the same entry or exit points by using this mobility-service. Nevertheless, they are driving together in a 'pooled solution' within one (possibly driverless) vehicle [28]. In terms of that, the occupancy-rate of vehicles seems to be raised within a city by using a pooled DRT-concept [29]. Thus, we state the hypothesis that the metric of vehicle miles travelled can be improved by a hybrid mobility system consisting of a pooled DRT-concept and strong public transport, that is accompanied by the previously introduced urban mobility hub concept at the respective intersections. A corresponding baseline improvement is the reduction of the number of vehicles which at least leads to a reduced necessity of parking space and thereby to the opportunity of a more livable design of the urban environment [30], as motivated before in Section 5.2.1.

A change in modality towards a pooled DRT-concept is based on multiple factors. A privately-owned vehicle has positive aspects like perceived costs, flexibility as well as comfort and privacy, while suffering from rising congestion levels on roads [31]. Therefore, a pooled DRT-concept has to significantly outshine privately-owned vehicles to enforce the described change of mobility paradigm.

5.3.1 Efforts of Operation and Usage for DRT-concepts and Hybrid Transport Systems

Mobility in urban areas is not yet for free, and organizations and users have substantial efforts when operating and using the ecosystem of mobility. Hence, the real costs of mobility are a huge underlying category when it comes to the final choice of modality within the urban environment. To address this, there is a need to take the operator's perspective to evaluate the whole cost of a given mobility service. Considering current research on this topic, it becomes obvious that there are several cost studies available for automated DRT-concepts that focus on ride-sharing instead of pooling solutions. While considering that the service of UberPool is cheaper (while adding possible detours) than the ride-sharing oriented UberX, we assume that the addition of automated driving technology to this comparison does not change the underlying price difference [32]. Thus, we analyze the cost factors of ride-sharing DRT-concepts in the following paragraphs to compare them with the costs of privately-owned MIV, while keeping in mind that pooled DRT-concepts are even less cost intensive.

Regarding a comprehensive view on the cost structure of automated ride-sharing DRT concepts [31] address the short-comings of previous studies and simulations, while taking into account, that the operating model of nowadays ride-sharing companies like Uber and Lyft is thereby to provide a digital mobility platform. Hence, on these platform solutions users can choose their respective kind of mobility service, while the actual vehicle operation is subcontracted to private drivers which offer their services on the platform. Towards this end, comparing the transport of only one passenger, automated taxis (urban: about 0.41 €/pkm, regional: about 0.32 €/pkm) outperform privately-owned automated MIV (urban, regional: about 0.45 €/pkm), when not only variable costs are included but also fixed costs [31]. Though, it is important to note that apart from the costs, automated privately-owned MIV can state a highly valued good among the population as much as specific vehicle brands nowadays, which state a symbol of freedom and individualism. At the same time there is a trend in vehicle sales that may counteract this effect. In particular, there are multiple brands like Volvo or Porsche offering their vehicles via a full-service subscription model on a monthly basis. These models render the full costs of a privately-owned MIV quite more realistically for the owner than the current variable costs most of the time perceived by users [33, 34, 35]. Focusing on the cost basis and leaving out these equalizing effects, the additional effect of increasing the mobility demand to more than one person per vehicle shifts the comparison even further towards the automated taxis. While there is a lot of research regarding the substitution of private vehicles or taxis with various concepts of automated taxis [36, 37] the opportunities of equipping conventional city busses with the automated technology, by the time when automated taxis are the state of the art, is often neglected. Estimated costs for such a city bus are halved to about 0.24 €/pkm for the urban environment, leading to the necessity of a more multimodal view [31]. Thus, against the background of the threat of rising congestion levels by a pooled DRT concept [30, 36], a hybrid transport system made up of a pooled DRT concept and public mass transport, consisting of automated city busses, trains and subways, is essential for the acceptance of this pooled DRT concept. Especially, in regard to congestion and to lower costs in comparison to privately-owned MIV [27, 30, 31, 38]. Summarizing, a pooled DRT-concept paired with a strong public mass transport system offers the same flexibility, lower costs and the same if not lower congestion levels in the urban environment than the current mobility paradigm focusing on privately-owned MIV. As stated earlier the urban mobility hub concept supports this hybrid mobility system while also leading to more flexibility because of its status as a new point of interest for daily life.

5.3.2 Necessary Changes to Automated DRT Vehicles

Next to the incentive of lower costs, a pooled DRT concept has to offer comparable comfort and privacy as well as personal space to a privately-owned vehicle [30]. When speaking about automated DRT vehicles there are multiple possible sizes ranging from solo vehicles up to vans with ten passenger seats, which lead to different base line vehicles for a pooled DRT concept. The necessary size of the vehicle depends highly on the mobility demand, the structure of the urban area, and the number of active vehicles in the respective urban mobility system, resulting in varying occupancies and waiting times for the passengers [28]. Assuming that an operator just offers one vehicle size and taking into account that from a financial perspective less vehicles are superior, while still being able to serve peak-hours, vehicles with a capacity of 8 to 10 passengers are the sweet-spot [28, 39, 40].

Comparing these vehicles to other forms of public transport, the same rules of personal space and privacy occur: An adjusted behavior occurs, which for example shows itself by avoiding opposing body orientations, because of the unavoidable invasion of personal space for a limited time period. This behavior though, does not prevent from an increased level of discomfort, so that design mechanism are necessary to prevent the invasion of personal space [41, 42]. Against the background of recently presented vehicle concepts and already active vehicles within ride-sharing systems such adjusted design mechanisms are not common practice [40]. The proposed solution of wide private seats with two armrests and a potential private table to increase the perceived personal space, would lead to high space requirements, limiting the number of seats to the lower end of the discussed spectrum of 8 to 10 people [43]. In addition, the level of discomfort can be reduced by giving the passengers control over their direct environment, in specific the lighting, temperature and seat position [44]. In conclusion, an 8-passenger sized automated DRT vehicle needs wide seats with sufficient boundaries like armrests or tables combined with active controls over the local passenger environment to achieve at least comparable comfort and a feeling of personal space to a privately-owned MIV. Regarding privacy, the link to personal space is quite close. In today's public space privacy has to be interpreted as the extent to which surrounding persons can have a look on your smartphone or laptop screen or whatever item one perceives as an object of personal information. Via the use of high chairbacks and individual visual covers between seats this issue can be addressed, concluding the mechanical design of an automated DRT vehicle.

5.4 Designing Mobility in Open Innovation Networks

The sole availability of a wide spectrum of current or prospective technologies and innovations, of which selected components have been presented in this chapters, are in their multitude not guaranteed to provide a path forward towards an improved mobility system. Instead of pursuing isolated solutions within their limited scope, the role of technologies and innovations as part of their integration into such complex ecosystems has to be comprehensively considered. In this regard, open innovation networks are presented to provide a basis for this multilateral realization, while incorporating the most relevant stakeholders, organizations and citizens.

5.4.1 Open Innovation Networks: Designing Ecosystems for Future Mobility

Despite the significance of the temporal mode in future urban mobility [45, 46], we still lack understanding towards the knowledge which practices and processes organizations use to handle the future challenges in mobility. In this context, one key activity in organizing things to come in future is to face digitalization [3]. Whereas the objective is clear that we will have digital processes of almost any kind in future, we hardly know how to engage in managing the digital transformation. Towards this end, one avenue that informs our reasoning and is important by designing ecosystems for the future is research on interorganizational networks [7, 8]. Based on a literature review, we found in our research that managing these constellations of three or more organizations is a common phenomenon across industries and fields as actors join networks for mutual benefits (e.g. sharing knowledge and resources). Hence, it seems to be difficult to compare empirical results in this realm due to the fragmented literature on this topic [7, 47]. Therefore, we first explain the multi-layered network concept and offer afterwards an informative account of how innovation networks in the mobility sector can be managed in an open and collaborative way.

5.4.2 Definiendum: Towards a Multi-layered Network Concept

Based on early contributions to the topic of managing networks by Moreno and Jennings [48], interorganizational collaborations are built upon the underlying phenomenon that at least three independent legal organizations are linked through relationships and keep up pace in terms of economies' scale and scope [7, 49]. There, activities carried out in this cooperative mode serve to pursue one or more common objectives and are usually coordinated in a cooperated way rather than competitive [47].

Towards this end, two levels of analysis can be distinguished by analyzing networks in general: Intraorganizational networks as purely intra-enterprise networking [50] and interorganizational relationships in terms of connecting organizations across their boundaries [51, 52]. As our literature review states, interorganizational networks are highly relevant in the field of mobility [46]. Hence, the reason for a collaborative and coordinated way beyond the boundaries of the organizations in the mobility sector is the demand for designing customized products and services, better quality, faster production and delivery times, lower costs and dealing with ever shorter innovation cycles. Finally, the main goal of all relevant stakeholders in this sector is to improve one's individual competitive position and to achieve efficiency and knowledge transfer in terms of managing interorganizational networks [52]. Therefore, we also focus on the level of interorganizational networks in terms of designing ecosystems for future urban mobility.

5.4.3 Building Open Innovation Networks

Globalization and technological change (e.g. digital products and services) are changing the boundaries between organizations and their environment in the mobility sector. Therefore, managing networks is not only a strategic but also an operational task. Structures and processes, but also the way to orchestrate organizations in a network, are changing from rigid and static ones to flexible-dynamic concepts. For this, interorganizational networks offer a high degree of flexibility and represent an alternative to vertical and horizontal integration [53].

Interorganizational networks in the sector of mobility aroused in terms of an adaption towards economic changes and digitalization [46]. This shift in the ecosystem of mobility forced organizations in this field of action to form multilateral networks to address customer needs and challenges such as risk reduction [7]. Organizations in these innovation networks are legally independent of each other but from the economic perspective they create an efficient information and production structure to gain synergies. Overall, the common goal is to reduce costs and increase flexibility through network exchanges but also to foster innovation and quality competition by opening company boundaries and, consequently, working together on ideas and solutions [52]. In the endeavor to achieve a deep understanding of future urban mobility it is highly relevant for organizations to identify structures, processes and the most important drivers that have an effect on the future mobility development in terms of managing the digital transformation. Towards this end, the network partners have to develop strategies and collaborate in an open way to build open innovation networks to address future mobility challenges [7].

5.4.4 Managing Open Innovation Networks

The coordination of and in networks is different to market or hierarchical structures due to the intermediary positioning. Interorganizational networks use both market control mechanisms (e.g. medium- to long-term pricing) and hierarchy (e.g. coordination of activities by a central actor). At the same time, networks are usually characterized by long-term agreements and a structure that has grown in the past [7]. Due to the predominantly existing heterogeneity of organizations in the mobility network, the entire network is often managed by a member which is active in the present network. Hence, the reason for selecting an organization to orchestrate the whole innovation network can be rooted in the value creation power or the reputation of the organization compared to the other actors in the network [52].

The management of interorganizational networks in the mobility industry is always goal-oriented in terms of the fact that control by instructions is not possible. Therefore, other control mechanisms are needed. Actors are independent of each other, so the managing organization always has the task to take care that all actors do not run out of the goals. As an underlying principle, open innovation networks offer a way for organizations to expand their own knowledge base and close knowledge gaps compared to the competitors [54]. The current discourse in literature of networks and open innovation shows that the existing innovation potential of innovation networks has not been discovered yet [55]. Previous innovation networks focus only on one aspect of the open innovation approach (e.g. the demand for an open exchange in terms of commercialization) while managing open innovation networks through an open exchange about organizational and network boundaries can lead to service and product innovations in the mobility sector.

As our findings illustrate, there is a gap in network research in terms of building and managing interorganizational networks in the field of mobility by an open way. Hence, we would like to close the gap with a practical contribution. On the one hand, we built an (open) innovation network in the area of Aachen, which is called "Living Lab Aachen" ("Erlebniswelt Mobilität Aachen") [56]. The network currently consists of 37 organizations (e.g. profit/non-profit, research or public entities) and deals with the topic of future urban mobility. Towards this end, there exist different goal- and topic-oriented think tanks to enable organizational openness through a multi-level exchange. Beyond that, these different think tanks act as a nucleus for innovations. The unique form of this network is the way it is managed. An impartial team orchestrates all these think tanks in a way, so that added value is achieved for all entities. On the other hand, out of the efforts of the local innovation network "Living Lab Aachen" and preliminary working groups, specific initiatives arise. These initiatives represent subgroups within the innovation network but can also integrate a significant number of further partners. Further, existing projects can be integrated with innovation networks in order to increase synergetic effects. Projects can manifest themselves multilaterally, or also as part of publicly funded projects. One example for this in terms of public funding is the project "APEROL". Based on previous findings, the project has the common goal to build a pilot operation of automated, electric moving vehicles in cities. Beyond that, these findings should, in a second step, be implemented in a comprehensive mobility system. Results and necessary synergies are fed back into the open innovation network. Through these activities, citizens and organizations should gain tailored mobility and transport services. To summarize, these two empirical examples show that a common goal is needed to manage open innovation networks in the mobility sector to finally design ecosystems for future mobility.

5.5   Conclusions

At first, we gave an insight on the rising challenges for urban areas all over the world, which they will face in the upcoming years due to the underlying megatrend of urbanization. To solve these challenges of urban mobility, a paradigm shift is necessary that is less focused on privately-owned MIV. As a solution we propose a hybrid mobility system consisting of a strong public transport, utilizing buses, subways and trains, and a pooled DRT concept, which bases on automated vehicles. Furthermore, the mobility system is accompanied by the concept of urban mobility hubs combining the park-and-ride concept with additional services at the point of transition between multiple modes like automated DRT concepts, rail and public transport. To compare the two paradigms the metrics of total costs, flexibility, congestion level and personal space as well as privacy were addressed.

On the mobility system level we found, that empowered public transport with automated busses as well as the pooled DRT concept with automated vehicles have less costs per person-kilometer than the use of an automated privately-owned MIV. At the same time the rising level of digitalization and user-oriented infrastructure leads to a stronger focus on intermodal mobility with comparable level of flexibility. Furthermore, based on multiple simulations of current research, a hybrid mobility system like the one proposed here, leads to lower levels of congestion in an urban environment.

On the vehicle concept level, we included the human factors that occur when fellow people are put together into a small space for mobility purpose. To reduce

Figure 5.2: e.GO Mover vehicle concept in urban scenario
(Copyright: e.GO MOOVE GmbH).

the related discomfort, adjustments to the mechanical design of the automated vehicle are necessary. For instance, to empower the feeling of available personal space two armrests per passenger as well as a space dependent small table are necessary. Based on this proposal and the results of several simulations from previous research, we concluded that a vehicle size of 8 passengers maximum is the sweet spot for a pooled DRT concept when also mobility demand, number of vehicles and expected waiting times are considered. Taking the psychological factor of direct control over the environment into account, several controls for temperature, seat position and lightning are also essential, next to the mechanical design. Finally, via the use of high chairbacks and individual visual covers, the matter of privacy has also been addressed.

Towards this end, our literature review in terms of open innovation networks in the sector of mobility showed, that the approach of interorganizational networks is highly relevant in this field. The reason behind this collaborative approach is driven by the demand for customized products and services as well as the overarching need for an ever-increasing individual competitive position. Hence, we focused on interorganizational networks in the context of future urban mobility.

Due to economic changes in the sector of mobility and the rise of digitalization, organizations from this sector were forced to focus more on customer needs and risk reduction. This led to a higher pressure to form multilateral networks. As part of the network, the legally independent organizations focus on synergies from efficient information, increased flexibility and common development of new ideas and solutions. In our analysis of this research area, we found a gap in regard to building and managing open interorganizational networks in the sector of mobility that we filled with two practical contributions.

The first contribution we presented is the development and ongoing management of the "Living Lab Aachen", an open innovation network of 37 partners consisting of profit/non-profit, research or public organizations, that all have an interest to shape the future of urban mobility. The key aspect in this regard is the impartial management of this open innovation network leading to a balanced added value for all participating organizations, and to provide a continuous integration into its processes. The second contribution results from the intense inter-organizational exchange within the think tanks of the "Living Lab Aachen", leading to multiple new initiatives and the integration of existing projects to leverage synergies in the sector of future urban mobility. For the domain of publicly funded consortia, one exemplary project is called "APEROL". It has the goal of implementing a pilot operation of an automated electric vehicle that embeds itself in a pooled DRT concept. The findings of this first step are later integrated into a comprehensive mobility system, which is backed by a city-based preceding traffic simulation. To summarize, our two practical contributions show the success of a common goal for the management of open innovation networks in the sector of mobility and for the design of the respective ecosystem.

References

1. March, J.G. 1995. The future, disposable organizations and the rigidities of imagination. *Organization*, 2(3/4), 427–440.
2. Beckert, J. 2016. Imagined Futures: Expectations and Capitalist Dynamics. Cambridge, Massachusetts: Harvard University Press.
3. Yoo, Y., R.J. Boland Jr., K. Lyytinen and A. Majchrzak. 2012. Organizing for innovation in the digitized world. *Organization Science*, 23(5), 1398–1408.
4. Bharadwaj, A., O.A. El Sawy, P.A. Pavlou and N. Venkatraman. 2013. Digital business strategy: Toward a next generation of insights. *MIS Quarterly*, 37(2), 471–482.
5. Pojani, D. and D. Stead. 2015. Sustainable urban transport in the developing world: Beyond megacities. *Sustainability*, 7(6), 7784–7805.
6. Yin, R.K. 2013. Case Study Research: Design and Methods. Thousand Oaks: Sage.
7. Sydow, J., E. Schüßler and G. Müller-Seitz. 2016. Managing Interorganizational Relations – Debates and Cases. London: Palgrave.
8. Kowalski. 2018. Management von Open-Innovation-Netzwerken. Wiesbaden: Springer Fachmedien, 2018.
9. United Nations. 2019. Department of Economics and Social Affairs, Population Division, World urbanization prospects: The 2018 revision. United Nations, New York.
10. Nobis, C. 2007. Multimodality: Facets and Causes of Sustainable Mobility Behavior. Transportation Research Record.
11. Gebhardt, L., D. Krajzewicz, R. Oostendorp, M. Goletz, K. Greger, M. Klötzke, P. Wagner and D. Heinrichs. 2016. Intermodal urban mobility: Users, uses, and use cases. *Transportation Research Procedia*, 1183–1192.
12. Monzón, A., S. Hernández and F. Di Ciommob. 2016. Efficient urban interchanges: The City-HUB model. *Transportation Research Procedia*, 1124–1133.
13. Parkhurst, G. 2000. Influence of bus-based park and ride facilities on users' car traffic. *Transport Policy*, 159–172.
14. Parkhurst, G. and S. Meek. 2014. The effectiveness of park-and-ride as a policy measure for more sustainable mobility. *Parking Issues and Policies*, 185–211.
15. Gallotti, R. and M. Barthelemy. 2014. Anatomy and efficiency of urban multimodal mobility. *Scientific Reports*, vol. 6911.
16. Lücken, V., N. Voss, J. Schreier, T. Baag, M. Gehring, M. Raschen, C. Lanius, R. Leupers and G. Ascheid. 2018. Density-based statistical clustering: Enabling sidefire ultrasonic traffic sensing in smart cities. *Journal of Advanced Transportation*. https://www.hindawi.com/journals/jat/2019/9317291/
17. Vlahogianni, E.I., M.G. Karlaftis and J.C. Golias. 2005. Optimized and meta-optimized neural networks for short-term traffic flow prediction: A genetic approach. *Transportation Research Part C: Emerging Technologies*, 13(3), 211–234.
18. Vlahogianni, E.I., M.G. Karlaftis and J.C. Golias. 2007. Spatio-temporal short-term urban traffic volume forecasting using genetically optimized modular networks. *Computer-Aided Civil and Infrastructure Engineering*, 5, 317–325.
19. Papa, E. and D. Lauwers. 2015. Smart mobility: Opportunity or threat to innovate places and cities. *In*: Proceedings of the 20th International Conference on Urban Planning, Regional Development and Information Society.
20. Harrison, C., B. Eckman, R. Hamilton, P. Hartswick, J. Kalagnanam, J. Paraszczak and P. Williams. 2010. Foundations for smarter cities. *IBM Journal of Research and Development*, 54(4), 1–16.

21. Bertolini, L. and F. le Clercq. 2003. Urban development without more mobility by car. Lessons from Amsterdam – A Multimodal Urban Region. *Environment and Planning A: Economy and Space*, 35(4), 575–589.

22. Louf, R. and M. Barthelemy. 2014. How congestion shapes cities: From mobility patterns to scaling. *Scientific Reports*, 4, 5561.

23. Çolak, S., A. Lima and M. González. 2016. Understanding congested travel in urban areas. *Nature Communications*, 7, 10793.

24. Litman, T. 2013. The new transportation planning paradigm. *ITE Journal*, 83(6), 20–27.

25. SAE International. 2018. SAE J3016 – Taxonomy and Definitions for Terms Related to Driving Automation Systems for On-Road Motor Vehicles.

26. Dargay, J., D. Gately and M. Sommer. 2007. Vehicle ownership and income growth, worldwide 1960-2030. *Energy Journal*, 28(4), 163–190.

27. International Transport Forum. 2015. Urban Mobility System Upgrade: How Shared Self-driving Cars could Change City Traffic. Paris.

28. Alonso-Mora, J., S. Samaranayake, A. Wallar, E. Frazzoli and D. Rus. 2017. High-capacity vehicle pooling and ride assignment. *Proceedings of the National Academy of Sciences*, 114(3), 462–467.

29. Henao, A. and W.E. Marshall. 2018. The impact of ride-hailing on vehicle miles travelled. *W.E. Transportation*. 46, 2173–2194. https://link.springer.com/article/10.1007/s11116-018-9923-2

30. Friedrich, M. and M. Hartl. 2016. MEGAFON – Modellergebnisse geteilter Fahrzeugflotten des öffentlichen Nahverkehrs. Final Report.

31. Bösch, P.M., F. Becker, H. Becker and K.W. Axhausen. 2018. Cost-based analysis of autonomous mobility services. *Transport Policy*, 64, 76–91.

32. Uber. 2019. UberPool. 04 November 2019. [Online]. Available: https://www.uber.com/de/en/ride/uberpool/.

33. Volvo. 2019. Care by Volvo. [Online]. Available: https://www.volvocars.com/de/carebyvolvo/.

34. Porsche. 2019. Porsche inFlow by Cluno. [Online]. Available: https://inflow.porsche.de/.

35. Gardner, B. and C. Abraham. 2007. What drives car use? A grounded theory analysis of commuters' reasons for driving. *Transportation Research Part F: Traffic Psychology and Behaviour*, 10(3), 187–200.

36. Fagnant, D.J. and K.M. Kockelman. 2018. Dynamic ride-sharing and fleet sizing for a system of shared autonomous vehicles in Austin, Texas. *Transportation*, 45.1, 143–158.

37. Bischoff, J. and M. Maciejewski. 2016. Simulation of city-wide replacement of private cars with autonomous taxis in Berlin. *Procedia Computer Science*, 83, 237–244.

38. Luo, S. and Y.M. Nie. 2019. Impact of ride-pooling on the nature of transit network design. *Transportation Research Part B: Methodological*, 129, 175–192.

39. Wang, B., S.A.O. Medina and P. Fourie. 2018. Simulation of autonomous transit on demand for fleet size and deployment strategy optimization. *Procedia Computer Science*, 130, 797–802.

40. Merat, N., R. Madigan and S. Nordhoff. 2016. Human-factors, user requirements, and user acceptance of ride-sharing in automated vehicles. *International Transport Forum Roundtable on Cooperative Mobility Systems and Automated Driving*, 6–7.

41. Bansal, P., K.M. Kockelman and A. Singh. 2016. Assessing public opinions of and interest in new vehicle technologies: An Austin perspective. *Transportation Research, Part C*, 1–14.

42. Evans, G.E. and R.E. Wener. 2007. Crowding and personal space invasion on the train: Please don't make me sit in the middle. *Journal of Environmental Psychology*, 27(1), 90–94.

43. Sanuinetti, A., K. Kurani and B. Ferguson. 2019. Vehicle Design may be Critical to Encourage Ride-pooling in Shared Automated Vehicles. Institute of Transportation Studies, UC Davis, Working Paper Series.

44. Averill, J.R. 1973. Personal control over aversive stimuli and its relationship to stress. *Psychological Bulletin*, 80(4).

45. Langley, A., C. Smallman, H. Tsoukas and A.H. Van de Ven. 2013. Process studies of change in organization and management: Unveiling temporality, activity and flow. *Academy of Management Journal*, 56(1), 1–13.

46. Spickermann, A., V. Grienitz and H.A. von der Gracht. 2014. Heading towards a multimodal city of the future: Multi-stakeholder scenarios for urban mobility. *Technological Forecasting and Social Change*, 89, 201–221.

47. Windeler, A. 2001. Unternehmungsnetzwerke: Konstitution und Strukturation. Wiesbaden: Springer-Verlag.

48. Moreno, J.L. and H.H. Jennings. 1938. Statistics of social configurations. *Sociometry*, 1(3/4), 342–374.

49. Chandler, A.D.J. 1990. Scale and Scope: The Dynamics of Industrial Capitalism. Cambridge, Massachusetts: Harvard University Press.

50. Tsai, W. 2001. Knowledge transfer in intraorganizational networks: Effects of network position and absorptive capacity on business unit innovation and performance. *Academy of Management Journal*, 44(5), 996–1004.

51. Gulati, R. and J.A. Nickerson. 2008. Interorganizational trust, governance choice, and exchange performance. *Organization Science*, 19(5), 688–708.

52. Provan, K.G., A. Fish and J. Sydow. 2007. Interorganizational networks at the network level: A review of the empirical literature on whole networks. *Journal of Management*, 33(3), 479–516.

53. Sydow, J. 1992. Strategische Netzwerke – Evolution und Organisation. Wiesbaden: Springer-Verlag.

54. Simard, C. and J. West. 2006. Knowledge networks and the geographic locus of innovation. *In*: Open Innovation: Researching a New Paradigm. Oxford, Oxford University Press, pp. 220–240.

55. Vanhaverbeke, W. and M. Cloodt. 2006. Open innovation in value networks. *In*: Open Innovation: Researching a New Paradigm, Oxford, Oxford University Press, pp. 258–281.

56. Erlebniswelt Mobilität Aachen. 2019. [Online]. Available: http://www.erlebniswelt.ac/ [Accessed 20 11 2019].

Communication and Service Aspects of Smart Mobility: Improving Security, Privacy and Efficiency of Mobility Services by Utilizing Distributed Ledger Technology

Ahmad Ayad, Johannes Borsch and Anke Schmeink

ISEK Teaching and Research Area, RWTH Aachen University,
D-52074 Aachen, Germany

6.1 Introduction

6.1.1 Motivation

Increasing urbanization and the associated new challenges are leading to a change in mobility. While there used to be plenty of room for individual activities and energy consumption, modern society demands a more efficient interaction in order to ensure greater road safety, smoother traffic and ecological sustainability. Innovative information and communication technologies make it possible to use existing transport infrastructures optimally and expand them with new mobility concepts such as car sharing, semi-autonomous, fully autonomous vehicles and new transport options, thus, intelligently improving traffic through new communication and service aspects. These new demands on mobility in modern society require new perspectives and approaches, such as environmental protection.

Smart mobility requires a high level of communication, both between vehicles (Vehicle-to-Vehicle, V2V) and between vehicles and the surrounding infrastructure (Vehicle-to-Roadside, V2R or Vehicle-to-X, V2X). This is only made possible by the significant advances in sensor technology, wireless

---
*Corresponding author:

communication, cloud computing and many more technologies used in IoT. These interesting technologies will enable devices to connect intelligently, transmit data ubiquitously and provide them with useful and necessary services. It will also require expanding the existing infrastructure and setting communication standards so that the cloud systems can be installed and communicate with a wide variety of IoT devices.

6.1.2 Problem Statement

The massive increase in such edge devices has led to a noticeable increase in the number of cyber-attacks on IoT devices. Since the primary purpose of IoT devices is usually not to communicate, but to perform their actual tasks, unlike computers and servers, their computing and power resources are severely limited, which is why traditional security measures often cannot be adequately applied. As a result, the often wireless communication can easily be manipulated and altered from the outside. In addition, the transferred data might include locations, personal or financial information, meaning there is a special interest in keeping it private and secure. In this respect, a new protocol is needed, especially for the Internet of Things, which helps to make the communication necessary for Smart Mobility to be more secure and ensures that there are standardized interfaces so that all possibilities can be exploited.

6.1.3 Outline

Although centralized security solutions are powerful and have an overview of the whole network, they introduce a single point of failure. For this reason, a closer look will be taken at the distributed security technologies such as distributed ledger technology (DLT) and their applications in IoT and Smart Mobility. In this context, a new communication system and protocol, called IOTA, designed specifically for IoT and based on DLT will be studied. Each IoT device acts as a separate node in the network, distributing information through the network and storing data about its neighbors. Thus, there are always some local copies that are used to confirm the authenticity of the received data and preserve the integrity of the transmitted information.

Since infrastructure alone is not yet intelligent, this chapter will further discuss applications and services that can make mobility smart such as a secure automatic tolling system and other secure traffic management systems.

6.2 Background and Related Work

As mentioned before, smart mobility will play a major role in the advancement of big cities in the future. With more people living together in a limited space, a more efficient traffic management needs to be created. This includes smart transportation, autonomous driving and driver assistance systems to increase road safety and improve ecological sustainability. To manage all these challenges, a reliable communication infrastructure is needed. One interesting approach to get this done is to use distributed ledger technology (DLT) as the underling communication technology between all the different devices.

6.2.1 Smart Mobility

Mobility today faces the challenge of becoming cheaper, cleaner and safer. At the same time the productivity and entertainment requirements in the car have increased. Laptops, tablets and smartphones are expected to be have access to information, goods and services anywhere, anytime. For smart mobility to keep up with all this, major changes and rethinking are needed to cause innovative information and communication technology breakthroughs. Foremost, it means that the individual components in the vehicle as well as the roadside infrastructure must be able to communicate efficiently and securely to support the driver in control and improve both the traffic safety and the ecological sustainability. This requires a tamper-proof communication protocol so that no malicious agent can infiltrate false information, which would cause dangers in road traffic.

Due to the enormous advances in communication and sensor technology and the associated growth in their complexity, cyber-physical systems have become increasingly vulnerable to cyber-attacks. This involves the correct functioning of the individual components. Also, it involves the reliable cooperation between the various relevant areas such as manufacturing, development and programming with the individual devices and the back-end services infrastructure, which is used to provide comprehensive functionalities or software updates. This entirety of smart mobility requires standardized interfaces, efficient management and secure communication. There is still much development to be done, for which DLT seems to be very suitable.

Specific initiatives in this area range from insurance systems to supply chain traceability, intelligent parking, vehicle rental and autonomous fleet management, all at different stages from market introduction, prototypes and concepts to visionary ideas. The application possibilities of DLT in the area of smart mobility can be divided into four fields:

- *Transport and logistics:* this field is driven by high competition and innovation. The focus is on process optimization in the cooperation between transport services and authorities.
  Currently, paper documents are still slowing down the process.
- *Mobility infrastructure:* This involves the availability and usability of charging stations for electric vehicles, vehicle licensing, online wallets, checkbooks or intelligent parking systems. These can be implemented in a decentralized and tamper-proof way using DLT-based protocols. The interesting part is that these functionalities can be used both with and without autonomous vehicles, i.e. they are already applicable and at the same time they accelerate autonomous mobility.
- *Mobility platforms:* In the future, different services are to be integrated

clearly and in a user- friendly manner into a uniform platform, for which DLT can offer a non-monopolistic, neutral basis. This could reduce complexity for users and overcome barriers to public transport use; thus, contributing to a more efficient road use and ecological sustainability. Above all, uniform standards are needed for implementable concepts.

- *Fully autonomous mobility:* At this point, DLT-based protocols can provide the key function of communication with the environment. Decentralization offers protection against malicious attacks or system failures. Intelligently controlled driving behavior can thus reduce fuel consumption and at the same time increase passenger safety, while fewer cars are required in urban areas and consequently fewer parking spots, creating space for a greater quality of life.

The next section will look in more detail at DLT, how far the technology has progressed, what are the benefits of using it and the current challenges that face it.

6.2.2 Distributed Ledger Technology

Distributed Ledger Technology (DLT) and especially its most famous representative, the Blockchain, known from Bitcoin, is becoming increasingly famous. It is already changing the financial market and the Internet. Additionally, it is able to change public administration in the medium to long term. According to [1], "a whole host of major financial companies have publicised their interest and investment in this breakthrough technology, and governments and international bodies are increasingly discussing the potential implications of distributed ledgers on business, government and the economy." In the following, the potential of the DLT, its applications and challenges will be discussed in more detail.

6.2.2.1 Definition

Jochen Metzger, director of the Bundesbank in Frankfurt, defines the term DLT in the Gabler Economic Encyclopedia as follows: "Distributed Ledger Technology (DLT) is a special form of electronic data processing and storage. A Distributed Ledger is a decentralized database that allows participants in a network a common read and write authorization. In contrast to a centrally managed database, this network does not require a central instance to create new entries in the database. New data records can be added at any time by the participants themselves. A subsequent update process ensures that all participants always have the latest version of the database at their disposal. A special form of the DLT is the blockchain" [2].

Accordingly, a distributed ledger is a database distributed over several locations, regions and/or participants. This means that, in contrast to a central ledger, all participants have access to the data records. The DLT overs a verifiable record of all information on all transactions. On the other hand, due to the absence of a central instance, all administrative tasks must be performed by the individual participants, so-called nodes, in the network. Thus, the transactions are processed and verified by the nodes. This creates a consensus about the truthfulness of the transactions and the current status of the entire ledger.

Although the technology was originally known for processing, verifying and storing financial transactions, the data records to be processed do not have to be restricted to such instructions, but can in principle contain any data and information.

6.2.2.2 Blockchain

The blockchain is known as the technology behind Bitcoin, probably the most popular crypto currency. It is a concrete realization of the distributed ledger technology and thus enables a completely decentralized system.

As the name suggests, the blockchain is a chain of blocks with time stamps. This chain is managed jointly by all participants in the network. Transactions are bundled in so-called blocks, and the blocks are then cryptographically linked to each other. To do this, each block is digitally signed and linked to the previous one by including the hash value of these blocks. New blocks can only be added at the end of the chain. Because of their mutual connection, existing transactions or blocks can no longer be individually updated, modified, manipulated or deleted. In this way, the block chain provides an unchangeable data memory. This main feature of the blockchain is the reason why transactions already executed can be trusted without having to rely on the evaluation of a central instance. Cryptography and digital signatures are used to prove identity and authenticity and correctly enforce read and write permissions to the blockchain.

The following advantages result from the characteristics mentioned so far and further goals connected with the technology:

- *Immutability:* Since all participants share the control over the system and all changes done to it, the only possibility for an attacker would be to change his forced transaction and directly all subsequent blocks at once; thereafter he would also have to convince all other nodes of his edits to the Blockchain. However, this is only possible if he has more than half the computing power in the network, an almost impossible condition. Thus, all already executed transactions are immutable.
- *Privacy:* Usually, participation in the network requires only a so-called private key and no further identification. This allows participants to be uniquely authenticated without having to draw any further logical conclusions about the person behind them. Obviously, this also brings certain risks. For example, Bitcoin is often used as a currency for ransom money in so- called ransomware attacks because the identity of the criminals remains secret. In addition, the loss of the private key is also accompanied by the loss of the entire invested assets.
- *Decentralization:* In contrast to conventional ledgers, where a central authority monitors the compliance with the rules and the balance sheets,

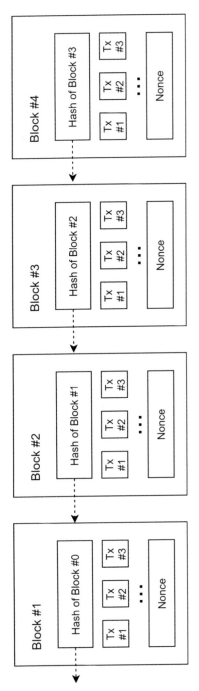

Figure 6.1: Blockchain.

Bitcoin's participants create a consensus themselves and control each other. In this way, everyone knows the current status and does not have to place their trust in a central authority.

- *Compatibility:* The goal is that different blockchains and databases with different contents can communicate with each other, so that data can be verified within a few seconds, without having to be recorded everywhere or to be collected from remote databases.
- *Transparency:* Distributed ledger technology allows large amounts of data to be made available to many users simultaneously. This makes it easy to track who provided what and when. This can be beneficial, for example, in the case of donations to allow traceability and check whether the money was really used for its intended purpose. Additionally, in the field of public administration, votes could be processed more efficiently, be more transparently documented and monitored.

Challenges: But before you praise the Blockchain and Bitcoin, you should try to understand their main features, because like every pioneer in a field, this technology is also struggling with some challenges.

According to the Bitcoin whitepaper, published in 2008 under the pseudonym Satoshi Nakamoto, transaction processing works as follows:

The individual monetary units are clearly traceable to prevent the so-called double-spending attacks. The earliest transaction is the valid one. In order to have a system that offers the participants a uniform history of the order, the Blockchain becomes involved. A transaction consists of the transaction data (value, etc.), the recipient's public key, and the hash of the previous transaction. Everything together is signed with the private key of the sender [3].

New transactions are distributed to all nodes. Each node collects new transactions in a one block. The node tries to find a difficult proof-of-work for its block. This proof-of-work serves as a consensus mechanism by which a new nonce is added to the block until the SHA-256 hash of the block begins with a specified number of zeros [4]. A nonce is an arbitrary number that only serves to vary the input of the cryptographic hash function in order to obtain the condition of the leading zeros in the output of the hash function with otherwise identical data in the block. Once this CPU effort has been completed as proof-of-work, the block cannot be changed without repeating the effort. Since later blocks are concatenated as in Fig. 6.1, the required work to change the block would involve re-executing all subsequent blocks, which is a significant effort. This CPU effort also determines the majority of the decisions when choosing the longest chain. As long as much of the CPU power is provided by honest nodes, the honest chain will grow faster and overtake everyone else [3].

If a node finds such a proof-of-work, it sends the complete block to all nodes. They only accept the block if all the transactions contained in it are valid and have not already been executed elsewhere. The nodes express their acceptance of the block by using the hash of the accepted block as the previous hash in the proof-of-work of the next block in the chain. The longest chain is always felt to be correct and will be extended [3, 5].

Since the entire history of the block chain of all nodes in the network must be stored locally, the nodes must have certain storage resources. To successfully solve the proof-of-work and be the first to do so, you need a very powerful CPU (or better yet GPU). These are the reasons why not every node that is interested in participating in the network, and using Bitcoin as a means of payment, is appropriate for these tasks. For this reason, there are two different types of nodes. One uses the technology merely as an anonymous and secure means of payment, while the other, known as full nodes or miners, store the complete history and compete for the fastest proof-of-work solution. To give these miners an incentive to invest computing time and power to advance the Blockchain, the creator of a block receives a certain amount of newly generated Bitcoin as well as fees for commissioning transactions. Overall, the following challenges can be formulated from the previous discussion:

Communication and Service Aspects of Smart Mobility: Improving Security, Privacy

- *Scalability:* Because a specific number of transactions are combined into a block (with size 1MB) and the proof-of-work solution takes a certain amount of time, the verification speed limit is currently around seven transactions per second. By comparison, other transaction service providers are much faster. For example, PayPal achieves 193 transactions per second and Visa can do up to 1,667 [6]. This is an unacceptable bottleneck for applications in the IoT area and thus Smart Mobility, where thousands or even millions of microtransactions must be executed.

- *Transaction fees:* Miners are necessary to keep the integrity of the blockchain. Transaction fees are essential to ensure that the considerable effort required is lucrative. However, in recent years, these have, in some cases, risen to several dollars per transaction. As a result, only transactions in which a high value is realized are worthwhile. Micro-transactions, such as those carried out between IoT devices, become unprofitable if the transferred value is a fraction of a cent.

- *Energy consumption and computing effort:* The computing effort required to solve the proof-of-work and effectively prevent manipulation is enormous. Bitcoin's estimated annual electricity consumption for 2019 is 73.12 TWh [7]. For comparison, if Bitcoin were a country, it would be in 40th place in terms of consumption, between Venezuela (75.2TWh) and Austria (72.2TWh).

- *Storage:* Since it is necessary to store the entire history of the blockchain locally on a node, a lot of storage space is required. At the end of June 2019, the entire blockchain had a size of approximately 226.6GB [8]. This size is not feasible for applications in the field of the IoT due the storage limitation of the nodes.

• *Conflict of interest between users and miners:* As there are different types of participants in the network, they also represent different interests. The problem is that they partly contradict each other. The main concern for users is to carry out transactions as quickly, securely, reliably and, last but not least, cost-effectively as possible. The miners, on the other hand, want to profit from their services, which is why they are interested in high transaction fees. In addition, they would like to have small and many blocks, which means that often new blocks are formed, resulting in frequent income. However, this also lowers throughput. Furthermore, only a few of the largest mining pools in the world have a large proportion of the computing power in the network. This means that if they cooperate, they could change the blockchain to their will, manipulate transactions arbitrarily and generate Bitcoin out of thin air without anyone being able to stop them. In the end, the idea of decentralization would be lost.

6.2.2.3 IOTA

As you can see, from the aforementioned points that blockchain oers inadequate solutions for some applications. Today, intelligent communication between devices via diverse networks is a fundamental base for smart mobility. Nevertheless, there are still very few global standardized protocols for the internet of things. According to [9], "In order to make such network function properly it needs convenient payment system and rapidly growing cryptocurrencies seem to be a natural candidate. [...] However, the vast majority of them are based on blockchain technology which has some flaws, especially noticeable within the IoT industry". Therefore, different groups try to create approaches for this platform that offer the advantages of the blockchain and find efficient solutions for its challenges. Probably the most promising approach in this respect comes from the IOTA Foundation, Germany's first crypto foundation.

IOTA calls itself "a permission less distributed ledger for a new economy. The first open-source distributed ledger that is being built to power the future of the Internet of Things with feeless microtransactions and data integrity for machines" [10]. IOTA is not an abbreviation; its a combination of the greek word (*iota*), meaning a small, insignificant quantity and IoT (internet of things). The name reflects IOTAs mission, which is to connect things using micro and/or zero value transactions.

Tangle: The so-called tangle is IOTA's distributed ledger technology and thus its alternative to the blockchain. The latter involves sequentially concatenating blocks that contain multiple transactions and are added at regular time intervals and sequentially referencing their chronological predecessors. In contrast, the tangle is not a chain, but more generally a Directed Acyclic Graph (DAG). The nodes in it are not formed by blocks, but by individual so-called sites, which contain only a single transaction. Each transaction refers to two previous transactions as shown in Fig. 6.2, which have already been verified. If there is no directed edge between transaction *A* and *F*, but there is a directed path with the length greater or equal than two from *A* to *F*, this means that transaction *A* indirectly verifies *F*.

At the beginning, all IOTAs, the tokens, were stored in one address and were sent by the first, so-called genesis transaction to several other founder addresses. This means that in contrast to Bitcoin, where new tokens are always created by mining blocks, all tokens ever available at IOTA existed right from the start. This genesis transaction is verified by all subsequent transactions, either directly or indirectly as shown in Fig. 6.2.

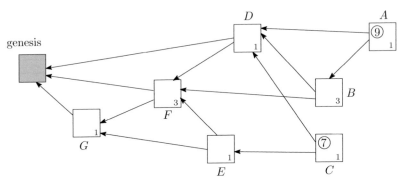

Figure 6.2: The tangle [11].

Unlike Bitcoin, at IOTA there is no difference between the nodes in the network. These nodes therefore process transactions and simultaneously verify others. This basic idea entails that everyone who wants to benefit from the network must also contribute to its security and integrity. A node for two theoretically arbitrary, still unconfirmed transactions, so-called tips, checks whether these are not in contradiction to each other and to the tangle history (e.g. double spending). If this is not the case, he attaches his transaction and automatically verifies his predecessors, otherwise he looks for others and does not approve the previous ones either directly or indirectly. In the whitepaper the mathematician behind IOTA and the founder responsible for the Tangle, Serguei Popov, explicitly points out that no rule is specified for the selection of the transactions to be confirmed [11]. Instead, he argues that if a large number of nodes follow a specific scheme, it is best for the other nodes to follow this scheme if they want their transactions to become confirmed. Moreover, to execute transactions, a node must do the following:

- Select two arbitrary tips.
- Check the tips for contradictions.
- Similar to Bitcoin, a cryptographic puzzle has to be solved. A nonce must be found so that the hash of transaction data and the nonce has a certain form. However, this proof-of-work is much more resource-efficient than Bitcoin's mining variant [12].

Another difference to Bitcoin is that transactions can be added to the tangle continuously and not in a discrete manner. As a result, the nodes generally do not see the same amount of transactions. Also, there is no longest chain that is recognized by everyone as the right one. Therefore, contradictory transactions remain in the ledger. Nodes decide individually which of the contradictory variants they confirm and which not. Because it is likely that a large number of nodes use similar schemes to be accepted as well, certain versions of the conflicting transactions orphan and are automatically filtered so that only one version survives. The orphaning of a branch in the DAG means that the transactions there will no longer be directly or indirectly verified by new transactions.

What motivates the nodes to select certain tips for verification and not to take random ones? One node assumes that other nodes use similar schemes. Then it calculates the probability by which different strands will keep getting verified based on its own theoretically arbitrary tip selection strategy. It selects these tips out of its own interest to increase the probability that its own transactions will be confirmed. If a node did so, it would be neglected by its neighbors.

What about manipulation and how quickly can transactions be varified? At the blockchain, the manipulation cost increases exponentially with the number of newly attached blocks to the block of the respective transaction. Also, for the tangle, the effort depends on the number of transactions verifying this transaction, although there is more than just one branch [3]. While the blockchain verifies transactions in relatively constant time intervals, the duration only depends on the difficulty of the proof-of-work. This can be dynamically adjusted in the number of leading zeros depending on the workload [12]. With the tangle, on the other hand, the number of parallel branches could grow arbitrarily or converge towards one. A divergence in the number of branches would imply that the number of tips and thus the number of unconfirmed transactions would increase enormously.

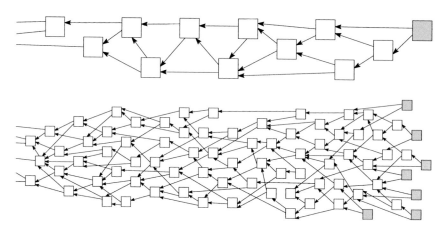

Figure 6.3: Low and high load [11].

This includes the result that the transactions could no longer protect each other from manipulation. If one follows the mathematical argumentation of the already mentioned white paper, one can assume that the number fluctuates around a constant value $L > 0$. Then one can distinguish two qualitative cases like the one in Fig. 6.3 as the following:

- *Low load:* L is small, often $L = 1$. This case occurs when only a few transactions are executed, i.e. it is unlikely that several transactions at the same time window verify the same tip. Even if the latency is low and/or the computing power of the participants is high, so that the execution time per transaction is very short, the load may be low. Of course, this only applies as long as no attacker artificially generates additional tips.
- *High load:* L is large. This can happen if many transactions are executed or otherwise the latency is high and/or the computing power of the devices is low, resulting in a long execution time. In these cases, it is likely that tips are selected for verification at the same time by many transactions (unknowingly) at once.

There is no clear boundary between these two extremes, but it is interesting to consider both. At low load with $L = 1$, the verification time is approximately only dependent on the rate of new incoming transactions. For high load $L \gg 1$, the verification time depends on the tip selection strategy. Thus on the execution time per transaction.

In any case, it is recommended to re-verify your own transaction by an additional no-value transaction, together with a promising tip, to promote it if it has not yet been verified after the double execution time. Since better transactions are preferred by other participants for verification, because they have a higher probability of not being orphaned. In contrast to the blockchain, which can only work with a low load, the tangle scales automatically as soon as the rate of incoming transactions increases. While with Bitcoin, the block size represents a natural bottleneck for the number of transactions per second, this speed is basically unlimited with the tangle as shown in Fig. 6.4. The more transactions are executed, the shorter the verification time for IOTA and the faster the effort to manipulate a transaction, i.e. its security, increases [3].

Network and Nodes: In contrast to a centralized network, where there is one server that controls all communication, or a decentralized network, where several servers control communication in different parts of the network, in a distributed network, all participants have the same role, the same rights and the same responsibilities [3]. Each network node is connected directly to one or more neighbors and indirectly to all other nodes via these neighbors. Thus, there are no hierarchically defined paths that messages in the network have to follow.

The IOTA reference implementation (IRI) is an open-source Java software that defines the IOTA protocol. Computers running IRI are called IRI nodes. They are the most important part of an IOTA network because they span the network

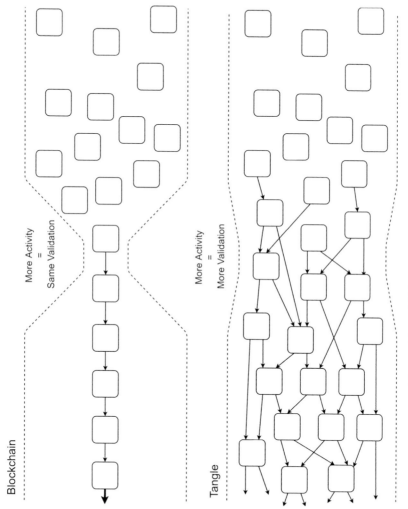

Figure 6.4: Scalability of blockchain vs. tangle [10].

and, according to the IOTA documentation, are responsible for the following key functions:

- Validate transactions
- Saving transactions in the ledger
- Have transactions executed by users and attach them to the ledger using the IRI node

To synchronize their ledgers with the rest of the network, all IRI nodes must send transactions to their neighbors and receive transactions from them. The main task is to validate transactions of neighboring nodes and then store them in their ledgers. If all local ledgers contain the same transactions, consensus is reached. This consensus forms the distributed ledger [10].

A complete IRI node requires a computer with at least 4 GB RAM and at least 8 GB storage for the tangle history, which is small compared to Bitcoin's requirements for miners, but usually still too large for applications in the IoT. For this reason, the so-called IOTA Controlled agenT (Ict) is currently being developed as a lightweight IOTA node especially for the IoT as an alternative to IRI [10]. The core functions of Ict consist, according to the official IOTA documentation, of:

- A communication protocol whose task is to send messages in distributed networks; these messages follow the typical IOTA transaction structure (i.e. validation of two tips, small proof-of-work, special signature); as with IRI, data, text or IOTA tokens can be transmitted.
- A modular approach called IOTA eXtension Interface (IXI). These IXI modules offer developers and users the ability to develop functionality independent of the basic architecture that uses Ict's protocol for fast, fee-free, distributed, and tamper-proof data transmission over the tangle, without getting too close to the realization of the actual communication.

In addition, the following theoretical functionalities are to be made available in the future:

- A swarm logic whose goal is to enable the collaboration of several small nodes in such a way that they act like a large node, thus enabling greater performance for complicated tasks.
- A consensus on the basis of economic clusters or groups, the so-called Economic Clustering (EC), which is to have the effect that small Ict nodes no longer have to make their decisions for the validation of other transactions on the basis of the entire tangles, but only refer to the events in the closer environment relevant to them, because only this influence on their actual field of application.

Vulnerabilities and Criticism: IOTA aims to achieve decentralization, immutability and trust without wasting resources in mining and transaction costs. However, some of IOTA's design features are different and it is uncertain if they will function well enough in practice. The following points summarize some of the issues and the criticism of IOTA:

- IOTA's data structures use a ternary implementation to improve efficiency; they uses trits (1, 0, 1) instead of bits. Because all the current hash algorithms are binary, a ternary hash design, called Curl-P, was implemented instead of well-studied alternatives that are used in other digital coins. However, cryptanalytic attacks on Curl-P proved to be possible [13].
- IOTA network is not yet mature enough as the number of participants and transactions is still low. Therefore, a central coordinator is used to prevent an attack on the IOTA tangle. This means that IOTA is not yet decentralized and can be shut down by a shutting down the coordinator [14].
- IOTA is considered safe against quantum computer attacks because of the Winternitz One Time Signature (WOTS) mechanism that it uses. However, spending from the same address several times significantly decreases the protection of the funds at that address, as it reveals portions of the private key associated with the address. This issue has caused funds to be stolen due to specific attacks [15].

6.2.3 Conclusion

In the previous sections, an overview of the communication and security related aspects in the area of smart mobility were discussed, along with the associated challenges. Also, the distributed ledger technology and its specific variants were explained. In addition to the classical solution of Bitcoin and the slightly different approach of IOTA's Tangle, there are many other concepts which use the distributed ledger technology for crypto currencies, but also more generally for secure data traffic, each with a focus on different problems. Like IOTA, others are also trying to better adapt to the limitations of the internet of things, but very few are doing so without blockchain. This creates a promising ecosystem for IOTA to cause a sensation in industry and business. In the next section, the possible applications of DLT in the area of smart mobility will be discussed in more detail.

6.3 Application Scenarios

The amount of data and financial transactions which smart mobility devices can generate is astounding. How to collect, evaluate and use this information in a secure manner to benefit people, cities and manufacturers is a critical issue in the industry today. Cars are being transformed into digital platforms. Paying for parking autonomously, tolls and other utilities. Distributed ledger technology provides a perfect platform for managing and securing transactions in smart mobility. However, managing transactions in a seamless manner is proven to be a challenge for conventional DLT technologies like blockchain. On the other hand, with IOTA, data and financial transactions generated by connected mobility can be processed securely, quickly, and without additional transaction fees.

In this section, we will explain in more detail how the distributed ledger technologies like IOTA can be applied in many areas. Examples include a proof-of-concept system, and an automatic toll system for individual calculation with tamper-proof data transmission, which was developed as part of a bachelor thesis [16].

6.3.1 Smart Tolling System

The current technology used for toll systems is outdated and inefficient, resulting in high costs and unfair pricing. Therefore, the basis and the concept for an automatic, intelligent and fair toll system has been developed here. It must be able to calculate individual prices based on various factors that are relevant for environmental protection and road use. These factors can include the average and maximum speed, vehicle size and weight, fuel consumption and emissions. In order to obtain these complex measured values, cooperation between the toll system and the vehicle is required. Since it is of interest that the collected data is calculated correctly and not manipulated by the vehicle owner in favor of a low fee, special attention must be paid to the tamperproof data storage and transmission. To guarantee this, IOTA's distributed ledger technology, the tangle, could be used as an interface.

The basis for the system is an IoT test network consisting of the following components and pictured in Fig. 6.5:

- *IoT nodes:* The IoT nodes represent the vehicles. They are the information sources in the network and generate the transmitted data. They are connected to the rest of the network via the MQTT protocol. Depending on the phase of the process, they send either authentication data or sensor readings to the toll system components. In the test network, temperature data is collected by a sensor integrated on the board and sent in order to test the data stream. In principle, however, the messages can contain any data or measured values about driving behavior, etc. In addition, each message contains a time stamp and an identification number.

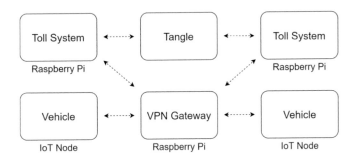

Figure 6.5: System overview.

The data itself is AES-CTR encrypted so that it cannot be read or manipulated by a third party.

- *VPN gateway:* The VPN gateway is implemented on a Raspberry Pi, which also acts as a VPN router, creating a VPN network for the IoT nodes and the other Raspberry Pis. Thus the IoT nodes remain anonymous to outsiders. The VPN Raspberry Pi is also necessary because the IoT nodes used are not able to connect directly to an enterprise network. By setting up its own network for the other devices the test network also becomes more portable. Should the application go beyond proof-of-concept, it would be replaced by other network technologies such as 5G.
- *MQTT broker:* The MQTT Broker is executed on a Raspberry Pi using the Eclipse Mosquitto™ MQTT Broker. The broker receives the messages published by the IoT nodes and forwards them to the recipients. (It is also possible for the Broker to decrypt the messages and store them in a local database).
- *Ict nodes:* The toll system itself is based on Raspberry Pis, on which the Iota Controlled agent is installed. The tangle, which serves for the transmission and intermediate storage of the data, is stretched between these. There are two different operating modes. If the test network is connected to the Internet, the Ict nodes can connect to other Ict nodes in the global tangle and use these for data transfer. For demonstration purposes or for portability reasons, however, it may also be useful not to connect the test network to the Internet, but only to establish a local network via the VPN router. In this case there is no access to the global tangle, so the data transfer only takes place between the Ict nodes on the Raspberry Pis.

The process flow is shown schematically in the following graphic (Fig. 6.6). When a vehicle reaches the toll road and stops in front of the barrier, the following process is triggered: First, the vehicle must identify itself to the toll system. To do this, it sends identification data to the toll station. The toll station checks the data and approves the continuation of the journey. The scheme of the received identity data is checked in the test system. As an extension there could be a database request in combination with the hash values in order to check if the account of the vehicle is sufficiently covered. From this point on, the vehicle periodically transmits the values measured by the sensors every 30 seconds to an Ict node, from here they are encrypted and sent through a channel, realized by an IXI- (IOTA eXtension interface)-module, on top of the tangle.

Until now, due to the limitations of the test system, only temperature, humidity and air quality have been dealt with. In a real system, this interface could be extended by including additional relevant data such as speed, fuel consumption and pollutant emissions. The other Ict nodes of the toll system also have access to this channel. As soon as the vehicle reaches the departure point, its identity is checked for the second time and the toll is immediately calculated individually based on the data stored in the channel. Due to the lack of reliable measured

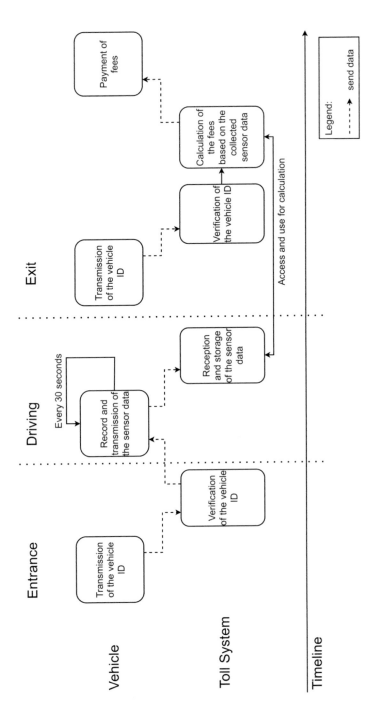

Figure 6.6: Process pipeline.

values, the fee has so far only been calculated based on the period of operation. The idea is to process payment quickly and securely with IOTA tokens, but it would also be conceivable to expand the interface to include standard payment services in order to allow all customers simple access. One passage ends with the payment.

6.3.2 Smart Cities

IOTA has been implemented in many smart city testbeds. An example is Powerhouse Brattørkaia in Norway. It is the latest and biggest energy-positive building in the country. The building will generate more power over its lifetime than it consumes and it incorporates smart wallet technology into vehicles that uses the IOTA Tangle distributed ledger. Using the wallet, a connected vehicle can both earn money and make micropayments for services.

Another application is the concept of a Dynamic Greenzone [17]. The emission values are to be collected using V2X communication, and dynamic green zones are to be determined based on the current pollutant load. These can then be automatically bypassed during navigation; alternatively, a fee similar to the toll system will be charged. This would flexibly and fairly avoid strict prohibitions, protect the environment and neighbors as well as reduce the load on traffic hubs during rush hour.

6.3.3 More on Smart Mobility

Another application is to transparently store tachometer readings and repairs in a distributed ledger. The stored information could also be signed. This would make tachometer manipulations almost impossible, given that they still occur far too often in the used car trade today. In addition, it would become clearly visible when which repairs were carried out and by whom (using the public signature). This makes it easier to assign responsibilities. Instead of complete data records, it would also be possible to store only hash values of the data records in the distributed ledger in order to ensure their integrity, even with lower storage requirements.

Finally, an innovative application in the area of V2V communication using the principle of platooning. The basic idea behind this is to save energy by taking advantage of the slipstream while driving. This seems obvious and relatively easy to implement, especially by trucks on long motorway drives. Additionally it could be worthwhile for other vehicles and in the city if the system functions reliably, as it also contributes to more efficient road use thanks to smaller safety distances. Of course, this requires hardware upgrades that go in the direction of autonomous driving, where only the leading vehicle steers and all others follow, without keeping the necessary safety distance for human reaction times. The leading vehicle benefits the least from the slipstream and so, to ensure equity, would need to be compensated accordingly. In contrast to central platforms, which can quickly lead to the formation of monopolies and the associated disadvantages, the DLT offers a neutral alternative. On the one hand, precise micro-payments can be made and on the other hand, data about the trip can be stored in the ledger, which makes it possible to draw conclusions about the cause of an error/accident, in the event that one occurs. A report commissioned by the German Federal Ministry of Transport and Digital Infrastructure also shows that drivers of rear vehicles may be able to have platooning counted as driver downtime in the future, which would also be documented in the ledger [18]. For this and the safety distances, there would still have to be legal changes. The report assumes that a nationwide use of platooning in Germany could save up to 500 million in fuel and 1.39 million tons of CO_2 each year. In the future, further savings in personnel and insurance costs as well as increased road safety are also conceivable.

6.4 Conclusion

Cheaper, cleaner, safer – vehicles are turning into rolling computers. Increasing urbanization will also increasingly lead to a change in mobility. Improved information and communication technologies make intelligent connections such as vehicle-to-roadside communication possible. New, universal standards and tamper-proof protocols are needed to safely expand the infrastructure and protect it against cyber-attacks. Decentralized solutions are ideal for this because they have no single point of failure. Distributed Ledger Technology is such a neutral instance. These decentralized databases, known through crypto currencies, stand for immutability, privacy, decentralization, compatibility and transparency. In practice, they face challenges such as scalability, transaction fees, energy consumption, computing effort, storage and conflicts of interest between different user groups. One approach that addresses the problems of the traditional blockchain is IOTAs Tangle, a directed acyclic graph (DAG). This basic idea carries that everyone who wants to benefit from the network must also contribute to its security and integrity. The tangle scales better because there is no limit by the block size. The more transactions are executed, the shorter is the verification time for IOTA and the faster increases the effort to manipulate a transaction (i.e. its security grows). At the end this chapter mentioned some use cases for DLT in Smart Mobility like an automatic toll system for individual calculation with tamper-proof data transmission as well as a dynamic greenzone application, the DLT as a transparent storage for tachometer readings and service intervals and the principle of platooning. DLT is able to verify identities through authentication and authorization, contributes to tamper-proof documentation and offers value transfers using tokens for billing purposes. For this goal, cross-application development to uniform interfaces and, of course, even more broadband expansion for better network coverage is required. These applications are still in early development stages or proof-of-concepts and need to be tested in sandbox environments to work properly, but they show the way to a connected, automated and universally mobile future that is not so far away and is necessary to handle some of the worlds current problems such as increasing urbanization and climate change through rethinking, greater efficiency and sustainable and security-conscious behavior.

References

1. McLean, S. and S. Deane-Johns. 2016. Demystifying blockchain and distributed ledger technology – hype or hero? *Computer Law Review International*, 17(4), 97–102.
2. Metzger, J. 2018. Distributed Ledger Technologie (DLT). February 2018.
3. Antonopoulos, A.M. 2017. Mastering Bitcoin: Programming the Open Blockchain. O'Reilly Media, Inc.
4. Nakamoto, S. et al. 2008. Bitcoin: A Peer-to-Peer Electronic Cash System.
5. Grinenko, M. 2020. Bitcoin and the Japanese Retail Investor. PhD thesis.
6. Vermeulen, J. 2017. Bitcoin and ethereum vs visa and paypal transactions per second. My Broadband.
7. Bitcoin energy consumption index, 2019.
8. Size of the bitcoin blockchain from 2010 to 2019, by quarter (in megabytes), 2019.
9. Kuśmierz, B. 2017. The First Glance at the Simulation of the Tangle: Discrete Model. https://www.iota.org/research/academic-papers, November 2017.
10. Iota website, 2019.
11. Popov, S. 2016. The Tangle. page 28. https://www.iota.org/research/academic-papers.
12. Franco, P. 2014. Understanding Bitcoin: Cryptography, Engineering and Economics. John Wiley & Sons.
13. Heilman, E., N. Narula, G. Tanzer, J. Lovejoy, M. Colavita, M. Virza and T. Dryja. 2019. Cryptanalysis of curl-p and other attacks on the iota cryptocurrency. Cryptology ePrint Archive, Report 2019/344, 2019. https://eprint.iacr.org/2019/344.
14. Muhammad Salek Ali, M., M. Vecchio, M. Pincheira, K. Dolui, F. Antonelli and M.H. Rehmani. 2018. Applications of blockchains in the internet of things: A comprehensive survey. *IEEE Communications Surveys & Tutorials*, 21(2), 1676–1717.
15. Sarfraz, U., M. Alam, S. Zeadally and A. Khan. 2019. Privacy aware iota ledger: Decentralized mixing and unlinkable iota transactions. *Computer Networks*, 148, 361–372.
16. Borsch, J. 2019. Distributed ledger technology for integrous and secure IOT communication. Bachelor thesis, RWTH Aachen. June 2019.
17. Radusch, I. 2019. Smart Mobility Services.
18. Gilbert Fridgen, G., N. Guggenberger, T. Hoeren, W. Prinz, N. Urbach, et al. 2019. Chancen und herausforderungen von dlt (blockchain) in mobilitat und logistik (Chances and challenges of DLT (blockchain) in mobility and logistics). May 2019.

Pseudonym Management through Blockchain: Cost-efficient Privacy Preservation on Smart Transportation

Shihan Bao[1], Yue Cao[2]*, Ao Lei[3], Haitham Cruickshank[1] and Zhili Sun[1]

[1] Institute of Communication Systems, University of Surrey, Surrey, U.K.
[2] Wuhan University, China
[3] Huawei Technologies, Beijing, China

7.1 Introduction

CYBER-PHYSICAL system (CPS) could be considered as one of the most promising techniques to improve the life quality of people. Here, one of the most attractive CPS cases is the Intelligent Transportation Systems (ITS), denoted as the Transportation-based Cyber-Physical System (TCPS). The combination of vehicle and network communication technologies has promoted the boundary of next generation, connected vehicles. This exerts pressure on car manufacturers to offer innovative products and services. While the connected vehicle and roadside infrastructure are physical entities, the Vehicular Communication System (VCS) is a network platform that provides Vehicle-to-Vehicle (V2V) and Vehicle-to-Infrastructure (V2I) communications. With the help of distributed computing infrastructures, the vehicle becomes a platform capable of receiving information from its peers and the environment, generating its own data, such as driver behaviour and car state, and transmitting data to other entities (e.g., vehicles, roadside infrastructure), in order to improve road safety, pollution control, insurance information and traffic efficiency.

In addition, the Internet of Things (IoT) technology [1] is driving traditional VCS towards the Internet of Vehicles (IoV). In paper [2], the authors believe applications in IoV rely on the exchange of Basic Safety Messages (BSMs) which contain vehicle status information such as location, speed, and vehicle dimension.

*Corresponding author: yue.cao@whu.edu.cn

As many applications and services make use of BSM, VCS faces the risks such as disclosing sensitive information about vehicles, adversarial manipulation of identity and location information etc.

The state-of-the-art schemes introduce pseudonym and pseudonym certificate to protect VCS security and privacy. Vehicles carry a set of pseudonyms which are used under different time periods in VCS communication. The safety beacon messages are sent using pseudonym certificates, instead of the long-term vehicle id (VID) issued by a certificate authority (CA). A set of pseudonym is a package including alternative ID called pseudonym ID, corresponding pseudonym certificate and corresponding public and private key pairs. Vehicles will sign theirs beacon messages using digital signature and attach pseudonym certificate before broadcasting messages.

Existing pseudonym and certificate management systems still left few challenges to overcome. A common solution called Security Credential Management System (SCMS) [3] has been well investigated, by providing a large scale system which can support 300 billion certificates per year for 300 million devices at full capacity. However, this advantage comes with a shortcoming: the system would have large certificate revocation lists and would be difficult and inefficient to achieve certificate revocation. The authors in [4] themselves illustrate the method of SCMS is prohibitively expensive regarding to storage limitations on the device (OnBoard Unit). This constitutes the motivation for the research we develop and report.

Pseudonym certificate management faces some additional challenges. Here, traditional network structures leads to ineffective creation of redundant pseudonyms, and it is hard to retrace malicious entities. Some studies [5] allow vehicles to carry a pool of pseudonyms, within which pseudonyms can be exchanged. Other studies [6] apply Road Side Units (RSUs) or certificate authorities (CA) to control the pseudonym exchange procedure. Since pseudonym related materials contain not only pseudonyms but also corresponding public/ private key pairs and certificates, it is expensive to generate and transmit massive amounts of pseudonyms. Furthermore, without suitable pseudonym management, all these pseudonyms would be used only for a limited time but will occupy the storage space of On-Board Units (OBU) in connected vehicles, even after their expiry.

In a traditional VCS structure, a central manager such as a Certificate Authority (CA) or Public Key Infrastructure (PKI) is designed to manage pseudonyms certificate centrally. However, a centralised network can be highly unstable, has low scalability, and represents a significant single point of attack. A number of pseudonym management schemes [7] state that a distributed and decentralised system could achieve better anonymity and durability. Since different locations would have different demands on pseudonym availability (based on the traffic and other factors), the assignment of pseudonyms is challenged by the variability of such needs. However, until now opinion suggests that decentralised RSUs appear to be unable to handle the pseudonym assignment problem efficiently (e.g., the paper [8] proposes a roadside unit (RSU) assisting pseudonym reused scheme using a distributed optimisation algorithm). Although the paper mentions the distributed optimisation algorithm, there is no fully explanation about how they fit that in their system.

With all this in mind, we envision that blockchain technology and distributed ledgers could be a feasible tool for resolving the challenges above. To tackle distribution optimisation problems in the shuffling process without a central manager, the pseudonym shuffling is realised by using the Blockchain distributed consensus. The pseudonym shuffling results are recorded in blocks (distributed ledger). The method also provides randomness of pseudonym shuffling and fully traceable record for certification revocation use. The blockchain technology brings robustness in the distributed structure. When a single point fails, the rest would still continue to work. The blockchain technology also provides randomness of pseudonym shuffling and fully traceable record for certification revocation use.

In this chapter, the proposed scheme aims for providing privacy-preserving pseudonym management that is more cost-effective across the system lifecycle than existing approaches. Firstly, a pseudonym Management scheme by using blockchain technology is proposed as the first contribution. Secondly, we introduce pseudonym certificate shuffling scheme, which is a new location privacy preservation scheme for VCS. It reduces pseudonym generation and management cost. A decentralised privacy manager (PM) is introduced in the system afterwards. The PM aims to alleviate the computation burden on RSUs and to improve the robustness of the network. As shown in Fig. 7.1, PM can be deployed as Multi-access Edge Computing (MEC) node within data network. The data gateway forwards data from MEC to the Radio Access Network (RAN, e.g. 5G base stations) and User Equipment (UE, e.g. mobile phones) access the MEC via the air interface between UE and RAN. Finally, asymmetric cryptography is used in blockchain transactions to protect pseudonym shuffle path. Each transaction is signed with sending PM's private key and encrypted by receiving PM's public key. As a result, other PMs in the blockchain network and attackers outside of the network cannot observe the information from this specific transaction.

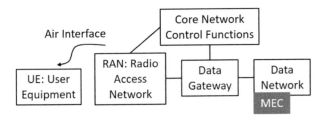

Figure 7.1: A brief 5G architecture.

The remainder of this chapter is organised as follows: Section 2 briefly introduces related techniques. The model overview and details of our scheme are discussed in section 3; we describe our system model, including the shuffling algorithms. The scenario for attack analysis and performance evaluation is given in Section 4. Section 5 concludes the paper and presents some plans for future work.

7.2　Location Privacy

7.2.1　An Introduction of Location Privacy

Despite the promises of VANET, researchers believe that vehicular networks will face huge challenges of privacy and security that might hinder the development of vehicular networks by users Raya [9] and Fonseca. [10] Recent revelations state that governmental programs globally collect, store, and analyse private data. Due to the broadcast feature of periodic message, many of these privacy violations are conducted by eavesdropping on communication networks. In addition, the vehicular communication primarily needs personal data including vehicular identity, position and direction. Even though it gives users plenty of benefits not only safety also entertainment, the privacy of vehicular networks has caused a massive concern. Malicious ownership of private data could allow attackers easily to explore individuals and to have the possibility to hack them. However, the trends indicate the future technology, not only in VANET but also in cloud services, will reply on high degrees of inter-connectivity, and analysing personal behaviour to provide more accurate customised services. Author in [11] states data will be inevitably abused. Such lack of privacy and sense of security deters individuals, especially drivers, from participating in these vehicular networks. As vehicular is the crucial part of this VANET project, the development of vehicular networks will face massive obstruction without support from drivers. Therefore, providing reliable, secure and private network services is extremely necessary.

In the vehicular networks, the privacy tends to be specific for hiding vehicular identity, position, and to prevent tracking by malicious entities. Speaking of vehicular privacy, there are five different properties defined by deploying Privacy-Enhancing Technologies (PETs) [12]: anonymity, unlinkability, undetectability, non-repudiation, and confidentiality. The five properties in vehicular networks situation will be explained respectively.

- Anonymity describes that a target vehicle cannot be distinguished from a set of other vehicles.
- Unlinkability describes that a target vehicle cannot be linked by malicious entity when there are two or more actions from the same vehicle.
- Undetectability means that the inability to allow unauthorised user discern whether certain message exists. Non-repudiation refers to the inability of a vehicle to deny its malicious or illegal behaviour.

- Confidentiality means that adversary did not have access to the content of data.

For vehicular networks, identity privacy of drivers and location privacy of vehicles are two crucial parts to be considered. Identity privacy requires a lot attention as it directly relates to individuals. But data life cycle management and database security, etc., are out of our scope. In this paper, location privacy is the main concern, which is hiding vehicular past, current and future whereabouts. Moreover, as the location privacy seems to be the most directly affected privacy domain, main properties investigated in this report are anonymity and unlinkability. Much research [13] pointed out that insufficient or absent measures to protect vehicular location privacy could cause a significant hindrance of drivers participating in VANET. Apart from the uncovering of personal and private data, there is the possibility to install extreme restrictive law enforcement systems. Therefore, drivers might not have a chance to choose whether they are willing to participate or not, as the systems will be installed compulsorily in the future.

More and more individuals seem to accept and tolerate the growing interest from industry, to collect personal information as they will obtain personalised service. Many online services which seem to be free of charge require the user to uncover personal data to be used. In this case, users are able to decide whether the benefits they obtained outweigh the value of information they abandoned. Therefore, some individuals [14] proposed that privacy or private information could be seen as a kind of currency.

In paper [15] the authors believe the location privacy is the ability to hinder third parties from obtaining the current location and location changes of vehicles. However, some studies [16] state that lots of test subjects could sell one month location data at only 35 for third parties commercial use. Many people do not strongly feel concern about the tracking of their location by a third party, while they feel mobile operators already collect their positions all the time. Hence, service providers suggest the preservation of location privacy might not be seen as a crucial part of the development of ITS. They believe that preservation of location privacy would not bring a significant impact on financial success of ITS. Furthermore, a business case about not preserving users privacy has been proposed as personal data which they believe can be used for targeted advertisement or sold to others. For instance, paper [17] believes many European mobile operators plan to sell customer location information.

So far, individuals usually have the right to choose if they want to keep their private information. However, in most cases, the scenario is quite different. The major benefits of ITS is traffic safety that is more important than their privacy. In the context of vehicular networks, drivers are forced to give up their location information when there is no protection of privacy. Even if some individuals choose their privacy over safety, future strategy seems most likely to leave them no choice as ITS service could be compulsory on the road.

A violation of location privacy could cause server effects. Exposed individuals not only could have obtrusive advertisements, but also have the implicit danger of being tracked by malicious third parties. Thus, unprotected vehicular location could cause serious results from embarrassment to financial loss or loss of social standing. Obviously, different entities who attack drivers location privacy can cause different levels of malicious influence. Those attackers could be other users of the same ITS system, such as service provider, application developer, system operator, or governmental department. Moreover, the violation of one privacy part could have influence on other parts. Especially, when location privacy is compromised, adversaries could use this information plan on malicious behaviour. For example, knowing an individual drove to the gym could indicate a certain period without further movements so that the adversary could plan attacks that might put the driver in danger. In order to prevent from these situations, a system must provide sufficient anonymity which is the precondition of location privacy. Anonymity means that an individual cannot be recognised and identified so that could preserve their location privacy. In the context of vehicular networks, there are many studies in terms of algorithms and schemes to provide location privacy. It would be important to compare which approaches could be realised, and to what extent the level of location privacy achieved in European and US is standard.

7.2.2 Blockchain and Blockchain-based Applications

In the past few years, Bitcoin [18] gains a lot of attention along with its blockchain concept, which was proposed in 2008. In simple terms, a blockchain is a synchronised and distributed ledger which stores a list of blocks. Each block records a set of validated transactions (e.g. user information and a receipt) and securely links to the previous block. The blockchain structure does not use central managers to control the system. It uses a digital public ledger to maintain and record runs by all participants nodes. This is realised by a protocol that achieves a trustworthy consensus about the chain of blocks created. In other words, network nodes can agree (deterministically) on the history and order of blocks that were created, and which node is allowed to add the next block to the chain.

The leader election of the node that can add the next block may be performed through a variety of techniques. For example, Proof of Work (PoW) poses a cryptographic puzzle to nodes based on a cryptographic hash function, the last local block seen, and the pool of transactions to be processed at a local node. A node that solves this puzzle announces the solution on the network, and other nodes accept such solution only if all transactions in the new block validate, the block does correctly point to the last block, and no other such solution was received beforehand. Since solving a puzzle is hard but verifying a solution is easy, this system provides security and effective validation and does not have a single point of failure.

The network will reach *eventual consistency* since some regions may temporarily diverge in their opinion of who won the next block. Since nodes hold an entire block *tree*, such disputes get resolved eventually as all nodes consider the path in the local tree with the "biggest overall work" to be the genuine chain (and this choice may vary over time).

- Blockchain offers a means of creating a trustworthy record of transaction histories in a network of nodes in which there exists mistrust. This is a conservative trust model for VCS, where some parts of the network would be within trusted computing bases (e.g. the CA) but other parts would be more open or even publicly accessible (e.g. the vehicles as nodes).
- Blockchain security is achieved in a manner that reflects the design choices of the blockchain. For example, when Proof of Work is used for consensus, the authors in paper [19] think one would need to control more than 50% of the nodes in the network in order to rewrite the blockchain history and so corrupt data veracity. This high degree of resiliency is what makes blockchains attractive in settings in which faults and malicious manipulation may corrupt integrity of data ledgers.
- Blockchains are beginning to be used not only for decentralised cryptocurrencies, but also for a wide range of applications including those in Internet-of-Things (IoT) scenarios, namely centrally banked cryptocurrencies [20], decentralising privacy [21], blockchain solution to automotive security and privacy [22], and proof of kernel work [23].

Despite the fact that blockchain has received a lot of attention from the banking industry, authors from [24] find that the use of blockchain can also improve other systems such as insurance, electric vehicles charging and car sharing services. The paper [25] states that there are some concerns about Blockchain, namely, majority attack, selfish mining, identity disclosure and abuse of Blockchain. In addition, blockchains based applications (e.g. Bitcoin and Ethereum) are facing big challenges in real life due to blockchain's low scalability and low amount of transactions per second. Hence, some studies from [26] proposed a new generation public digital ledger technique, namely IOTA [27]. It makes use of a small notation called "TANGLE" at its core, a guided acyclic graph (DAG), to eliminate huge transaction power consumption and the concept of mining. Blockchain technology can be used for certificate management system in IoT and ITS. The paper [28] is proposed as an extension for pseudonym certificate management based on the decentralised pseudonym management system to complete the lifespan of the certificate. It focuses on the certificate revocation system using the similar decentralised blockchain system of the chapter to prevent from insider attacks and reduce the communication overhead.

7.2.3 Privacy Attacks in IoV

In paper [28] a survey comprehensively analyses security and privacy requirements in vehicular networks. Privacy threats were studied and classified into the following categories of attack. The Trace Analysis Attack is used for tracking a mobile phone. The previous cloaking areas are linked to the movement pattern of the user. In paper [29], a location-based system (LBS) server can derive the chances of the mobile user being at different locations of the cloaked area. In paper [30], bogus location proofs are generated when two nodes collude with each other. For example, if a malicious node m_1 needs to assert that it is in a location at which it is not, it can have another compromised node m_2 to mutually generate fake location assertions to prove m_1 is in that location.

The authors in [31] propose the Trajectory Attack used against users location privacy. The attacker could analyse the target's previous locations to conduct a trajectory remap. In paper [32] the authors state that trajectory attacks are possible even if the identifier of the user has been removed.

Attacks on data integrity will fail to provide the trusted services to the users and vehicles. The authors in [33] have evaluated the attacks on data integrity on real-time traffic information manipulation that is generated and passed by the vehicles in the ITS. The paper states that data integrity attacks could disrupt the ITS service information and even cause a severe traffic congestion.

In paper [34] the authors claim that the Transition Attack means that the attackers make use of previous observations data for transition prediction at junctions. In paper [35], the authors believe that the actual movement of the target can be reconstructed by the attackers (using possible trajectories of users based on the probabilities to events). Similar to trace analysis attacks, adversaries monitor historical movements for target future location predictions.

7.3 Proposed Framework

7.3.1 System Model

Nodes in VCS are hierarchically classified into four layers, based on their responsibilities. There are three layers for the service providers, while the service user occupies a single layer. As shown in Fig. 7.2, the service provider

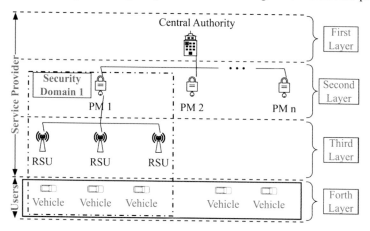

Figure 7.2: VCS network hierarchy [28].

comprises RSUs, PMs and Public Key Infrastructure (PKI). The PMs and RSUs have wireless communication devices which can communicate over the wireless medium, utilising VCS communication standards (DSRC [2, 36] or/and C-ITS [37]). Here, RSUs act as access points (APs) which offer interfaces to bridge messages between the service provider and users. Moreover, we assume that each vehicle is required to be equipped with a built-in computerised device known as an *On-Board Unit* (OBU), in order to support the VCS standards. A PKI contains a Certificate Authority (CA), an Anonymity Server (AS) and other third-party infrastructure that may support applications.

All the pseudonym-related cryptographic materials, such as anonymous credentials, key pairs and pseudonym certificates are created by the PKI. Each PM has its own logical coverage area, called the *security domain*. PMs help the PKI to manage cryptographic material of security domains that are logically placed below the PKI layer. Privacy managers are placed in a geographically sparse manner. Each privacy manager covers one security domain managing a number of RSUs in its coverage. With equipping the on board unit on the vehicle, vehicles are able to broadcast and receive basic safety messages with other vehicles and RSUs on the road. A basic safety message consists of a pseudonym, a timestamp, and the current vehicle status information, such as speed, directions, geographical positions, and vehicle dimensions.

Vehicles equip a set of pseudonym certificates which are used under different time periods in VCS communications. For privacy protection purpose, vehicles should only use every pseudonym Id and corresponding certificate for a short period. The valid duration and change scheme for pseudonyms has been recommended by different entities. While the American VCS standard SAE J2735 [38] asking vehicle to change pseudonym within 120 seconds period or 1 kilometres travel, European standards ETSI TS in ETSI TS 102.867 [39] recommends vehicles to change pseudonyms in every 5 minutes. The RSUs are equipped with the same network communication technology, and are fixed infrastructures with a certain communication coverage area (e.g., a radius of 300 meters in DSRC protocols). The RSUs relay messages between vehicles and PMs, which act as service providers of VCS. To provide context, we compare the traditional and blockchain-based network structures.

7.3.1.1 Traditional Network Structure

The traditional structure is based on the hierarchy structure. As shown in Fig. 7.3(a), the network is based on a hierarchy structure. On the top level, public key infrastructures maintain and control the system centrally. On the middle level, different security domains consists of privacy manager and RSUs. Privacy managers are relays for PKI and local RSUs, in order to pass security and privacy management information efficiently. Each PKI manages several PMs, as many as are appropriate for the geographical topology of the area. Moreover, PKIs act as bridges that connect different security domains.

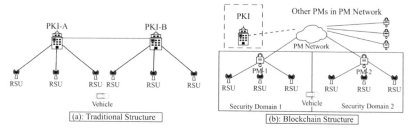

PM: Privacy Manager   PKI: Public Key Infrastructure

Figure 7.3: Network structures: (a) Traditional (b) Blockchain-based [28].

Inspired by the previous work [40] we introduce PMs to cover the privacy-related function of VCS. The PM can be seen as the Security Manager (SM) in the previous work, which has extended privacy protection functions. The RSU is a stationary device placed along roads and at intersections, which is used to gather information about the road traffic and broadcasts it to the OBUs that are within communication range. Also, an RSU can communicate with other RSUs and the CA to exchange messages related to the road traffic through a secure channel. Our previous work [41] follows the traditional network structure as do most other works.

7.3.1.2 Blockchain-based Structure

The PMs manage a certain amount of RSUs based on the geographic distribution of RSUs, shown in Fig. 7.3(b). In contrast to a traditional network structure, a PKI is isolated and would be a part of an existing authority such as a Driver and Vehicle Licensing Agency. The PKI is designed to generate specific cryptographic credentials for all legitimate nodes and to map vehicles credentials to their long-term identities. In paper [42], the authors state that cryptographic credentials for ITS environment commonly contains vehicle long term identity, alternative id pseudonyms, corresponding cryptographic key pairs, and related pseudonym certificates. All crypto materials should be stored in a secured and trusted entity to preserve privacy and security. As a result, the central authority will have access to the system in only the two cases:

(i) *Initial registration:* When fresh new vehicles come out from manufacturer factory, CA will generate long term vehicle IDs and the first batch of crypto materials for vehicles.

(ii) *Adversary revocation:* After the initial malicious behaviour detection mechanism starts and accountability reports deliver to CA, the CA would retrieve information from the blockchain system. That is when CA needs full access to the system. In the proposed blockchain-based structure, blockchain look-up is used to identify the malicious entity and to achieve repudiation.

As a result, the proposed blockchain-based structure could enable PMs to securely keep all communication logs without reliance on a central party. All PMs are connected with each other and the PKI on a domain. PMs communication mainly contain peer-to-peer pseudonym sets exchange, encapsulated in transactions. Similar to Bitcoin, the ledger keeps all transactions from the beginning. And PMs act as miners to put transactions into a block within a fixed period of time. With this blockchain-based structure, our system can reuse pseudonyms by shuffling them between PMs. The shuffle results will be determined by the first miner and be added to the block. Hence a blockchain can be maintained for the purposes of pseudonym management. We also made assumptions for the blockchain structure:

Assumption 1: Role of Miners

Generally speaking, nodes are classified into two roles according to different responsibilities among the blockchain network, namely service user and miners. The miners are nodes with powerful computation power who use their computation power to maintain the blockchain. In the Bitcoin network, nodes decide on their own whether or not they want to take on the role of a miner. Bitcoin pays the miner who wins the mining race for the next block a reward, in addition to transaction fees embedded in that block. This creates incentives that ensure that mining takes place, but also causes problems such as dramatic increases in difficulty when Bitcoins become very valuable in fiat currencies.

In the blockchain-based scheme, we assume all the block mining tasks are carried out by all the PMs as procured resources, and so they do not need any incentives and won't necessarily receive rewards – as discussed in [43] previously. This is sensible in our setting because we believe that pseudonym management, as part of ITS management, should be run by the appropriate organisation of the government (e.g. the Driver and Vehicle Licensing Agency in the UK). All the PMs take the roles of service user and miner at the same time. It may also be attractive to use Proof of Kernel Mode [44] as a variant of Proof of Work that randomly and securely would select an expected number of PMs for mining each time. This will then allow for using a lower level of difficulty and will save costs as only those nodes selection for the next mining race will consume energy in mining.

Assumption 2: Approximate Mining Synchrony

It is beneficial to be able to ensure that all the PMs start mining tasks at approximately the same time. As the navigation service is contained in the ITS applications, each vehicle should have a synchronised clock. This helps to limit the deadline for each transaction collection interval. Any lack of synchrony may also be contained by using a combination of, for example, Proof of (Kernel) Work and Proof of Elapsed Time.

Assumption 3: Consensus

Proof of Work is the only consensus mechanism that has been tested successfully and in a sustained manner in a highly adversarial environment, and is the only known cryptographic puzzle that meets these testing requirements. Alternative consensus mechanisms such as the ones aforementioned have not yet been tested in real and adversarial practice. This is why we favour PoW-style consensus given that an ITS is part of a regional or national critical infrastructure that may be subject to aggressive attacks, perhaps even facilitated by compromised insiders. PoW gives us this resiliency even against corrupted PMs and low levels of difficultly, especially when used with a Proof of Kernel version of PoW, which give a balance of security and shorter compute time at lower cost.

7.3.2 Threat Model

Due to the nature of broadcast safety messages, an eavesdropper may monitor a specific vehicle movement and track its location information by eavesdropping periodic basic safety messages broadcasting from the target vehicle. In this chapter, we consider external and internal attacks. The two types of external attacks are *global passive attack* and *local passive attack*. The two types of internal attacks are *internal tricking attack* and *internal betrayal attack*.

Global Passive Adversary (GPA): A global adversary has the overall coverage of a connected vehicle network. The global passive adversary could eavesdrop broadcast beacon messages on the road so that is able to monitor any vehicle location in any region of interest.

Local Passive Adversary (LPA): The LPA is opposite to GPA. Unlike the global passive adversary, the local passive adversary could only eavesdrop limited coverage which based on the radio transmission range.

Internal Betrayal Adversary (IBA): IBA is one of the internal attacks, which mainly focuses on a previous legit node being compromised and still staying in the system as a legit node. So it could spoof critical safety related messages and even collaborate with outside global attackers to attack target vehicles.

Internal Tricking Adversary (ITA): The internal tricking adversary will use pseudonyms which have been allocated to others, allowing it to confuse the vehicular network system and to attack other nodes. ITA stays internal and will use other vehicles pseudonyms or used pseudonyms to make the current ITS bewildered.

There are other methods for attacking the vehicular network system. For example, accessing traffic monitoring cameras or hijacking the Global Positioning System (GPS) allows tracking the target vehicle. Furthermore, adversaries may be able to compromise privacy managers to attach a false block into the blockchain. Yet acquiring either of these capabilities, access to a traffic monitor that controls the national traffic operations centre or taking control the blockchain itself, requires a significant effort – e.g. having at least 51% of the total blockchain network's processing power.

7.3.3 Pseudonym Management

We now introduce the blockchain-based pseudonym management scheme, which intends to reuse pseudonyms and address the distribution issue that decentralised systems have regarding pseudonym shuffling. The main symbols used in this scheme are listed in Table 7.1.

Table 7.1: Symbols used in the paper

Notation	Description
PM_x	The xth PM.
PKx, SK_x	The public and private keys for the vehicle x.
PN_i^{PMx}	The ith pseudonym from PM_x.
PM_x	The xth PM.
n_T	the number of transactions.
$\mathbb{R}$	a set of all possible number of transactions.
cipher in fo	Encrypted information of pseudonym sets with public key of destination PM.
$Sig\{CI\}_{SK_{PM-s}}$	Digital signature on CipherInfo with private key of the source PM.

7.3.3.1 Pseudonym Distribution

From a management perspective, pseudonym sets for each car that are presently stored in the OBU will be depleted. Authors in [45] mentioned that the use of a backbone network of RSUs may resolve the aforementioned issues and reuse pseudonyms for a limited period of time and in different geographic areas. However, the distributed nature of these systems then creates an additional optimisation problem: it is hard to balance the volume of incoming and outgoing pseudonyms without a centralised means of controlling this. This issue remains even when using distributed versions of the simplex algorithm in order to alleviate the computational demands on the optimisation problem.

In terms of Privacy-by-Design for the VCS network, we should consider pseudonym generation and distribution more wisely. Specifically, the number of pseudonyms generated by a PKI should be limited but sufficiently large in order to meet demand. Two initialisation events are introduced to finish the entire initialisation stage, namely the permanent identity and pseudonym generation. The permanent identity contains the identity number **ID**, a certificate **CERT** and key pairs (private key **SK** and public key **PK**) which are used to prove the real node identity or the initial registration identity. These credentials are generated by PKIs and distributed to the manufacturers who are responsible for producing vehicles and the VCS infrastructure.

The distribution procedure between a PKI and manufacturers is finalised via highly secured connections, such as optical fibre or cable connections. PKIs generate a certain number of pseudonyms offline and then distribute pseudonym sets to each PM. Each pseudonym set $\{id_1 \cdots id_n\}$ contains the corresponding pseudonym certificates $\{cert_1 \cdots cert_n\}$ and encryption private/public key pairs $\{sk_1/pk_1 \cdots sk_n/pk_n\}$. The number of pseudonyms inside sets is determined by the density of traffic in corresponding areas which all RSUs reported to their

PMs. Pseudonyms will be encrypted and signed to maintain secrecy before a PKI distributes them to PMs. The encryption and signing use the public key PK_{PM} of the PM and secret key SK_{PKI} of the PKI, respectively. To summarize:

i) *Generates Permanent Identity:*

PKI *generates **ID, CERT, SK & PK***

ii) *Distributes Permanent Identity:*

PKI *sends **ID, CERT, SK & PK** to* **Manufacturers:**

$\{ID+CERT+SK+PK\}_{secured\ channel}$

Manufacturers *issues **ID, CERT, SK & PK** to* **Vehicles**

or **Infrastructures:**

$\{ID+CERT+SK+PK\}_{file\ transfer}$

iii) *Generates Pseudonyms:*

PKI *generates:* $\{id_1 \cdots id_n\}$, $\{cert_1 \cdots cert_n\}$ *and* $\{sk_1/pk_1 \cdots sk_n/pk_n\}$

iv) *Distributes Pseudonyms:*

PKI *sends pseudonyms to* **PM:**

$\{id_1 \cdots id_n\}_{PK_{PM}} + \{cert_1 \cdots cert_n\}_{PK_{PM}} +$

$\{sk_1/pk_1 \cdots sk_n/pk_n\}_{PK_{PM}} + \mathbf{Signature}_{SK_{PKI}}$

7.3.3.2 Pseudonym Shuffling

To keep sufficient pseudonyms to allow frequent changing across vehicles, PMs are responsible for retrieving used pseudonyms and issuing fresh pseudonyms. There are two challenges for this shuffling scheme: (1) the path of pseudonym exchanges needs to be protected; otherwise, the attacker could subject the path to further analysis in order to constrain the possible pseudonyms delivered to vehicles in certain RSUs' ranges. (2) The demand of pseudonym for each privacy manager is supposed to be fulfilled. For instance, the PM covers central London would need more pseudonyms than PMs in countryside because different locations have different traffic. We use blockchain technology to deal with these challenges for pseudonym management, as it could provide sufficient randomness on the shuffling path and enough computation power to tackle the distribution optimisation problem.

When vehicles operate on a road, they will frequently change pseudonyms based on a certain pseudonym change algorithm. For our pseudonym management, pseudonym changes are supposed to execute within mixed zones, which are geographic regions within the VCS environment as shown in Fig. 7.4. Generally speaking, the mixed zone must be selected carefully in order to maximise the level of privacy. Places where many vehicles tend to gather around, for instance road junctions and roundabouts could improve the level of privacy of vehicles.

Algorithm 1 briefly describes the mechanism used when a vehicle joins a mixed zone. We propose that traffic junctions and roundabouts could be treated

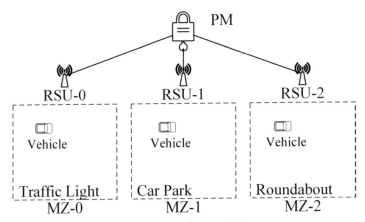

MZ: Mixed Zone   PM: Privacy Manager

Figure 7.4: Mixed zone example.

as physical mixed zone where RSUs can be placed, while traffic lights or other places at which enough vehicles could gather in close proximity are seen as "virtual mixed zones". In virtual mixed zones, vehicles would trigger pseudonym change even when the vehicles are on a highway with vehicles of a similar status (e.g., similar speed, same heading, and so forth). A vehicle first hides its location information according to the specific cloaking algorithm in the mixed zone. This aims to gather all vehicles so that the chance of the target vehicle being tracked can be minimised.

Algorithm 1: The Joining-Mixed-Zone Mechanism

Input: Current PM id PM_x, Public Key of PM_x: PK_x, used pseudonym set PN_{used}, a **Mixed Zone** area under managed by PM_x, Location Cloaking Requirement of **Mixed Zone**: $Cloak\{\}$, Current location *Location*

1: **if** (Vehicle enters a RSU cover area) **then**
2:     **Mixed Zone = True**
3: **else if** (Vehicle enters a virtual mixed zone) **then**
4:     **Mixed Zone = True**
5: **else**
6:     **Mixed Zone = False**
7: **end if**
8: **if (Mixed Zone = True) then**
9:     Cloaks the location information $Cloak\{Location\}$;
10:     Broadcasts safety messages using cloaked location;
11:     Encrypts pseudonym by PM's public key: $Enc\{PN_{used}\}_{PK_x}$;
12:     Sends $Enc\{PN_{used}\}_{PK_x}$ to PM_x;
13: **end if**
14: **End Algorithm**

Initially, vehicles carry a set of pseudonyms installed at the time of vehicle manufacture. A vehicle marks a pseudonym as "**used**" and switches to a new one if the pseudonym meets its expiry conditions. We defined a threshold for used pseudonym sets, a fixed percent of the number of pseudonyms.

To assure the vehicle will not only have enough new pseudonyms to use after it gave up used pseudonyms but that it also collects a maximum number of pseudonyms in order to reduce transmission overhead, the threshold is set to cover the majority of pseudonyms. The vehicle then encapsulates all used pseudonyms into a package that it sends to the current RSU. PMs will collect used pseudonym packages for a fixed period of time from all RSUs that are situated in its range and then aggregate all packages into a single transaction. All packages and transactions are signed by their senders and encrypted with the receiving PMs' public keys. Hence PMs could assure all pseudonyms are integrated and authenticated. Then the PMs upload all used pseudonyms, related indexes, and the number of pseudonyms in the *PM cloud*. After that, each PM in the network will make a copy of all pseudonyms that have been uploaded for this shuffle. Since every communication between PMs contains timestamps, pseudonym shuffle will be triggered in every fixed interval. When pseudonym shuffling commences, all PMs pull the demand of each PM from the cloud and add those demands to their own list. The PMs will randomly choose pseudonym sets and allocate them to every PM based on the number of required pseudonym sets.

Algorithm 2: The Pseudonym Shuffling Scheme

1:   **for** $(x = 1; x \leq i; x++)$ **do**
2:       **PM$_x$** gathers all the used pseudonyms from mixed zones it manages;
3:       **PN^{PMx}** $= \{PN_1^{PMx} \cdots PN_{nx}^{PMx}\}$;
4:       Counts the number of used pseudonyms $= n_x$;
5:       Encapsulates **PN^{PMx}** into package and sends into **PM** cloud network;
6:   **end for**
7:   **for** $(x = 1; x \leq i; x++)$ **do**
8:       **PM$_x$** picks up all the pseudonym package within PM cloud network;
9:       **Shuffles** the pseudonym sequence and **relocates** to destination PMs;
10: **end for**
11: All the PMs start **Mining**;
12: The mining winner broadcasts the **Block** into PM network;
13: **for** $(x = 1; x \leq i; x++)$ **do**
14:      Retrieves new pseudonyms for **PM$_x$**;
15: **end for**
16: **End Algorithm**

The shuffling algorithm is outlined in Algorithm 2. Table 7.2 illustrates an example of all the forwarded packages within the PM cloud network of *i* many PMs, ranging from PM_1 to PM_i. The first field in the package header indicates the type of this packet, used for further extending the service to security applications.

The remaining fields in the header are the PM number $\{n_1 \cdots n_i\}$ and the number of pseudonyms which are donated from the PM, respectively. The payload field contains all the used pseudonyms PN^{PM}.

Table 7.2: The format of forwarded package

Package header			Payload
Type	PM no.	PN number	Pseudonyms
Privacy	PM_1	n_1	$\{PN_1^{PM_1}, PN_2^{PM_1} \cdots PN_{n_1}^{PM_1}\}$
Privacy	PM_2	n_2	$\{PN_2^{PM_2}, PN_2^{PM_2} \cdots PN_{n_2}^{PM_2}\}$
...	...	...	...
Privacy	PM_i	n_i	$\{PN_1^{PM_i}, PN_2^{PM_i} \cdots PN_{n_i}^{PM_i}\}$

An example of the shuffle mapping result is shown in Table 7.2. Here we assume random variables $\{a, b, c, d\} \in [1, i]$ and $aa \in [1, n_1]$, $bb \in [1, n_2]$, $cc \in [1, n_3]$, $dd \in [1, n_4]$.

The first line means a PM selects a previously-used pseudonym $PN_{aa}^{PM_a}$ from PM_a. This source to destination result is marked by the sequence number 1. After creating this list, each PM encapsulates its list into a Blockchain-based transaction. Then PMs will try to mine for consensus, e.g. by calculating Proof of Work (PoW). Whoever first finishes that mining race must add the mined block into the Blockchain. All PMs will validate such a new block and, if validated, follow the block's description of how to allocate pseudonym sets. Due to the nature of encrypted messages, the transaction would only be feasible to specific sender and receiver of that transaction. Even though all transactions are attached

Table 7.3: Shuffle mapping table

Src	Dest	Seq no.	Pseudonyms
PM_a	PM_1	1	$PN_{aa}^{PM_a}$
...	PM_1	...	...
PM_b	PM_1	n_1	$PN_{bb}^{PM_b}$
...	...	...	...
PM_c	PM_i	$\sum_{k=1}^{i} n_k - n_i + 1$	$PN_{cc}^{PM_c}$
...	PM_i	...	...
PM_d	PM_i	$\sum_{k=1}^{i} n_k$	$PN_{dd}^{PM_d}$

in the block and the block was broadcast to all PMs, others who neither sent nor received a specific transaction cannot obtain any information from that transaction. Hence each PM can only decrypt its own transactions (for which they were the receiver). So each PM will perform the pseudonym shuffle by shuffling all pseudonym indexes individually. After each shuffling, all PMs will delete all copies of pseudonym sets. The format of transactions and mining will be described next.

7.3.3.3 Transaction Format

The transaction ledger is designed to encapsulate pseudonym materials from a source-privacy manager to a destination-privacy manager. An example of the transaction ledger is shown in Table 7.4. In the ledger, the left-hand side shows the source PM address and the destination PM address, while the right-hand side includes the pseudonym sets of this transaction and their corresponding credentials materials (e.g., key pairs and certificate). In the transaction payload, data information will be encrypted using Elliptic Curve Integrated Encryption scheme (ECIES) with receiver's public key. So it provides message integrity and confidentiality. In addition, the transaction will be signed using Elliptic Curve Digital Signature Algorithm (ECDSA) with sender private key, which provides messages repudiation and prevent from impersonate attack.

Table 7.4 shows the format of each transaction in the ledger, containing transaction header and payload. In the transaction header, the number of this transaction specifies the position at which this transaction is located in the ledger. The source and destination PM address are similar to Bitcoin inputs and outputs seen in paper [46]. In order to keep the integrity, authenticity and non-repudiation, last field of transaction is the signature. The *cipher info* has already been discussed above.

Table 7.4: The format of transaction

Transaction header
Hashed result of the transaction
Number of this transaction in block
Current privacy manager address *PM-sour*
Destination privacy manager address *PM-dest*
Signature of this transaction to ensure integrity and authentication
$Sig\{Cipher\ Info + PMdest\}_{SK_{PM\text{-}sour}}$
Payload: (Encrypted Transaction Information)
$Cipher\ info = En\{Pseudonym\ sets\}_{PK_{PM\text{-}dest}}$

7.3.3.4 Block Format

A block is designed to store all transactions as ledgers. All blocks need to be joined to form a large chain—the *blockchain*. The format of a block is shown in

Table 7.5. The first row shows the block number, which is the sequence number of the block within the entire chain. The hash of the previous block securely links this block to its parent one through the mining process. This makes is extremely hard to replace contiguous sub-chains with other data and to convince other nodes of the validity of such changes. The Merkle tree root [47] is a well known technique for keeping data integrity. In the proposed case, each transaction in one block will be authenticated into the Merkle tree root. As a result, the system will notice any alteration on transactions since it would change the Merkle root value. Similar to Bitcoin application, the proposed scheme adds the timestamp into the block to prevent from time tampering attack. The fields for targeted difficulty and nonce are designed for Proof of Work, which creates a digital receipt of which first node mined that block. The mining process and Proof of Work for the proposed approach is described in the next section. The payload field contains the aforementioned transactions that the block creator randomly allocated.

Table 7.5: The format of block

Block header	
Field	Description
Version	Block Version Number
Previous Block Hash	Hash of the previous block in the chain
Merkle Tree Root	Hash of the merkle tree root $Root_M$
Timestamp	Creation time of this block
Targeted Difficulty	The Proof-Of-Work difficulty target
Nonce	A counter for the Proof-Of-Work
Block Payload (Transactions)	
Transaction No.1 . . . Transaction No.n	

7.3.3.5 Consensus Algorithm

The consensus mechanism used in a blockchain establishes, in a distributed way, an agreement between all network nodes, instead of relying on a central party's decision. The most widely known and used consensus mechanism for blockchains is Proof of Work (PoW), which is a mining race in which nodes try to solve a hard cryptographic puzzle concurrently. In paper [48], the PoW system was originally proposed as a means of deterring spam email. All PoW-based applications (e.g. Bitcoin and the current Ethereum) require participating nodes to contribute a significant amount of computation power in order to obtain a digital proof of work that can be verified easily. The process by which nodes compete in finding such proof of work is called *mining*. The first node to solve the cryptographic puzzle for the next block to be added to the chain will be elected/accepted as leader and is then able to add the new block for which it found proof of work to the chain.

In blockchain-based applications, a cryptographically strong hash function is used to calculate the proof of work. In our case, we use double SHA-256 on the previous block hash result and the Merkle tree with related time stamp of the new block as input to that hash function.

The proof of work involves adding some random information, a nonce value, to that input until the resulting hash has a desired minimal number of leading 0 bits. Consequently, proof of work has several desirable features. For example, a miner that has had k failed mining attempts has no advantage in the $k + 1^{th}$ attempt in comparison to another miner who just begins its first attempt of solving PoW. Also, PMs are very likely to have different mining times due to the exponential distribution for the expected time to find proof of work within a certain period of time. Also, we assume that all PMs randomly generate transactions that result in different root hashes and that they all use the same hardware specification, making this a blockchain system in which miners are procured resources as proposed in paper [49]. Therefore, all PMs have the same probability of getting the correct hash results within a certain period of time t, assuring the randomness of the resulting shuffling.

Since all PMs contain identical processing modules and since they are assumed to link with highly secured wire connections, we may set the level of difficulty (the number of leading 0 bits in the hash output) required to be rather low. This low level of difficulty allows for a short Proof of Work computation time, resulting in an efficient yet resilient consensus mechanism.

7.3.3.6 Shuffle Time Composition

Table 7.6 demonstrates each time factor for the shuffling process. The variable t_{prep} is the time needed for preparing a block, including the PM's generation of a

Table 7.6: The time elements of processing procedures50

Parent field	Description of parent field	Child field	Description of child field
		t_{rand}	Calculation time to generate random transactions
t_{prep}	The time cost to prepare block which will be mined later	t_{fill}	Time cost to insert transactions into the block message
		t_{merkle}	Calculation time to get Merkle Tree Root
		t_{header}	Processing time to prepare block header
$t_{transfer}$	Transmission time cost in PM network	t_P	Propagation time in network cable
$t_{processing}$	Processing time for message **Verification**	t_V	Processing time to verify signature
t_{mining}	Mining time for shuffle process	t_M	The average time to mining a block

randomised transaction ledger and the time cost of block preparation. We denote by N the number of PMs. To calculate the total time cost of the shuffling process, we also need the number of transactions ($n_T \in \mathbb{R}$) where $\mathbb{R}$ is a set of all possible number of transactions. Given that,

$$\mathbb{R} = 2 \times (n - 1) \times n \qquad (1)$$

where $n \in \mathbb{Z}^*$ and $\mathbb{Z}^* = \{0\} \cup \mathbb{Z}^+$. $\mathbb{Z}^+$ denotes the positive integers. Hence the total time cost can be described as:

$$t_B = n_T \times t_V + 2 \times t_P + t_{prep} + t_M \qquad (2)$$

So the total time cost can be expressed as seen in equation (2), that contains all time factors. Note that the total transaction verification time ($n_T \times t_V$) depends on the number of transactions (n_T). The preparation time and mining time are added to reflect the time needed for creating a block in the blockchain.

7.4 System Evaluation

7.4.1 Privacy Attack and Defence Analysis

Researchers such as Yu in [51] state that the power of identity and location privacy preservation in pseudonym-based systems is determined by the unpredictability of mapping temporary identifiers (pseudonyms) to vehicular permanent identities. So the proposed blockchain based pseudonym management system is aim to provide unlinkability and unpredictability of pseudonym change process. In addition, it reduces the management cost regarding to a large-scale networking system. Privacy attacks pose a serious issue in current ITS that require addressing. Without proper identity and location-privacy preservation, attacks, such as vehicle tracking, location manipulating and so forth could cause serious damage to vehicles and compromise the safety of human actors. Moreover, the lack of such abilities will hinder the development and acceptability of the Internet of Connected Vehicles.

In the following, we show how our approach can address some pertinent attacks and defence measures.

7.4.1.1 GPA and LPA

The most common privacy attack is when an adversary passively eavesdrops vehicles' beacon messages. GPA has bigger coverage compared with LPA. Passive adversary tends to accumulate information about surrounding vehicles in terms of location and timestamps from broadcast messages. Both GPA and LPA are able to use those information to derive the possible location remapping. As mentioned in Section 2, several works claim that they can predict vehicles' trajectories with a brute-force collection of beacon messages even when vehicles change pseudonyms frequently. In contrast, our proposed system not only allows vehicles to change pseudonyms simultaneously at a mixed zone, but also at the virtual mixed zone as long as there are sufficiently many vehicles with similar context within that zone. In this case, for both GPA and LPA, the unpredictability of mapping vehicles is accumulated.

7.4.1.2 IBA and ITA

As stated above, the internal betrayal adversary and internal tricking adversary will be considered as the insider attackers in this case. Both IBA and ITA are categorised into internal attacks. The IBA could use target vehicle's existing pseudonym to impersonate the target and also collude with outside GPA to form a trajectory of the target vehicle. In contrast, the proposed scheme prevents the IBA from accessing others' pseudonyms simply, as vehicles will not exchange pseudonyms with each other. According to the pseudonym change scheme of the proposed system, vehicles only update pseudonyms from their own pseudonym sets which have been allocated by the RSUs (not vehicles). After acquiring a new pseudonym set from the RSU, the vehicle cannot retrieve the original source of the new pseudonyms in that set. Therefore, the IBA could not obtain any useful information from its surrounding or related vehicles.

In terms of ITA, the malicious user will keep and repetitively use pseudonyms that have been uploaded to RSUs and allocated to other vehicles. While other vehicles are using the same ones, the ITA could use the pseudonyms to confuse the vehicular network system and launch other attacks. To deal with this problem, the system behaves as follows. If the adversary stays in the current RSU's coverage, the RSU will realise that the adversary keeps using the old pseudonyms. Then the RSU will mark the vehicle as adversarial and broadcast this information to other vehicles. If the adversary leaves the RSU's coverage, no one other than a CA could know that the attacker is a ITA. That is because the feature of the blockchain based shuffling system: in the block, all transactions are encrypted using different corresponding receivers public keys. Only sender and receiver of each transaction can access the transaction which includes related pseudonym shuffle plan. So RSUs and PMs will not know the allocation of each pseudonym and also could not decide whether the adversary is using false pseudonyms. However, once the attacker launches other attacks the false pseudonyms, such as spoofing messages, that can be detected, the CA will retrieve transactions of the blockchain. Hence it can spot the adversary and perform revocation of original credentials.

7.4.1.3 Compromised PM

The privacy manager is a crucial part of this pseudonym-management system. Therefore, we normally assume that PMs are relatively secure, most likely run and maintained by government agencies or similar governing bodies. In addition, blockchains are well known for providing high robustness and for being hard to manipulate by an adversary. However, let us consider a worse circumstance in which one PM is compromised by a malicious user. There are several attack scenarios that may be enabled by this. However when a PM is compromised or has lost connection with the blockchain network, the whole blockchain will discard the PM after repeated failed attempts to acquire its response. In addition, all pseudonym sets that this PM received from previous round pseudonym shuffles will be abandoned from the PKIs and will thus not be used again.

7.4.1.4 Spoofing Block Attack

The spoof block attack assumes that a privacy manager (PM) has been compromised or betrayed so that it is broadcasting false blocks into the blockchain cloud. Then an adversary can re-arrange pseudonym allocations and manipulate vehicles' identity. But in order to consistently send out forged blocks and to have them accepted by the blockchain network, the compromised PM will need to have at least 51% computation power of the total blockchain based network since Proof of Work is used as consensus mechanism. As a result, it is extremely difficult for attackers to control the whole mining process since the massive energy consumption cost.

7.4.2 Performance Analysis

7.4.2.1 Simulation Assumptions

OMNET++ with Veins and PREXT [52] packages were used for the simulation work. Both Elliptic Curve Integrated Encryption scheme (ECIES) and Elliptic Curve Digital Signature Algorithm (ECDSA) were used for security encryption and digital signature purpose. We now offer a quantitative analysis of our approach. The simulation of the proposed scheme was carried out using OMNET++ with the dedicated simulation package (Veins and PREXT) The Elliptic Curve Integrated Encryption scheme (ECIES) is used for encryption scheme, while the Elliptic Curve Digital Signature Algorithm (ECDSA) is designated for the digital signature. We simulated the proposed scheme on our desktop machine equipped with an Intel Core i7, 8 GB RAM, and a display card Inter HD Graphics 530. Our simulation considered 300 vehicles. All vehicles followed the pseudonym change scheme of our proposed system. We set the traffic density at 50 vehicles per kilometre and the transmission range of the target at 50 metres based on [53, 54]. The performance results are broken into four parts. Firstly, we evaluate the pseudonym reuse frequency. Secondly, we calculate the total amount pseudonym usage and compare with EU ETSI standard. Then the degree of anonymity is studied compared to other existing schemes. Lastly, we investigate the time cost of the entire process.

7.4.2.2 Pseudonym Reuse Frequency

We first study the pseudonym reuse frequency to demonstrate shuffling effectiveness. We let each vehicle carry 10 pseudonyms with a threshold setting of 8, meaning whenever a vehicle has 2 unused pseudonyms it will upload 8 used pseudonyms to its RSU. We simulate the scenarios in which the shuffling process happens 50, 100 and 200 times. Figure 7.5 shows the percentage of how many pseudonyms have been reused over 1, 2 and 3 times respectively. As it can be observed from the result, over 60% of the pseudonyms have been used at least twice while 200 shuffling process. Furthermore, the frequency of reused pseudonym decreases when the number of used pseudonyms are getting bigger. Hence the results of pseudonym reuse frequency indicate that our scheme assures pseudonyms could be reused in different locations in limited shuffling iterations. Despite the fact that the increased reuse times of each pseudonym could free up more storage for OBU, the results also show that the number of pseudonym reuse times affects the percentage of reused pseudonyms. Only 0.15% of the pseudonyms have been reused over 3 times in 50 shuffling iterations. Therefore, we need to have a reasonable understanding of the suitable number of shuffling times used to measure the anonymity performance of our scheme.

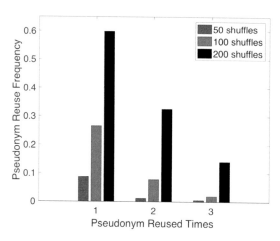

Figure 7.5: The percentage of pseudonyms re-usage vs. pseudonym reused times.

7.4.2.3 Pseudonym Total Amount

To quantify the efficiency of reusing pseudonyms, we compare the total amount of pseudonyms that the proposed system needs for a 24-hour period with the ETSI standard [55] for the change frequency. Since the ETSI standard suggests vehicles change pseudonyms every 5 minutes, one vehicle needs 288 pseudonyms for 24 hours. For the proposed scheme, we denote that the capacity of storing pseudonyms of each vehicle is X. Based on the size of storage in OBU, vehicles currently could have from 10 pseudonyms to nearly 1,000 pseudonyms. We continue use 100 vehicles in this comparison. The system following the ETSI standard would need over $288 \times 100 = 28,800$ pseudonym certificates each day and 864,000 each month, whereas the proposed scheme uses 100,000 pseudonyms even with a storage capacity of 1,000 pseudonym certificates until the system decides to replace new pseudonyms. In fact, our scheme is not affected by time duration and pseudonym certifites would be re-used over time.

7.4.2.4 Anonymity Set Size

Based on the result of frequency of reuse pseudonym, the simulation was carried out ruining 200 times of shuffle process. Figure 7.6 shows the impact of the k neighbours for the expected the anonymity set size. The anonymity set size is a measure of the level of anonymity.

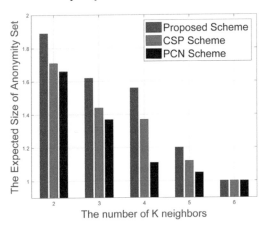

Figure 7.6: Expected anonymity set size with different value of k neighbours.

Note that the relative anonymity level cannot go below 1, as 1 means one vehicle exists alone. Higher size of anonymity set means that messages are anonymous and hided with larger k values than the user-specified minimum k-anonymity levels. From the result, the proposed scheme has outperformed the benchmarks which are coordinated silent period (CSP) scheme [56] and the cooperative pseudonym change (PCN) scheme Pan2012. The CSP proposes an approach that all vehicles in certain area completely cease any communication and changes its pseudonym for a period of time to maximise the anonymity. The PCN illustrates that the target wait till k neighbours around to change pseudonyms together. The proposed scheme achieves a better level of anonymity as the expected anonymity set size is greater. In addition, all schemes from the simulation have drops in terms of the expected AS size during k getting bigger. Because it is less probable to find greater or identical k neighbouring vehicles when the value of k becomes large. When k is set to 6, the anonymity set sizes of all three schemes are equal to 1, which means that vehicles can only find less than k neighbours.

7.4.2.5 Processing Time

We illustrate the total processing time of our proposed scheme. We acquire each time component from Table 7.6 and calculate the result of total time cost based on equation (2). As can be seen in Fig. 7.7, the total time increases when the transaction number grows. Due to the benefits that come with our design of privacy managers, the transaction number is limited as equation (1) demonstrated. Therefore, the total shuffle process time stays within a reasonably short period. For instance, we take a medium size city as an example. We assume 30 privacy managers are placed in the city, and each of them covers several RSUs. Based on equation (1), the number of transactions can be less than 100 in off-peak hours, while the maximum number could exceed 1000 in peak hours. The total processing time varies from 0.2 seconds for 100 transactions to over 2 seconds for 1,000 transactions.

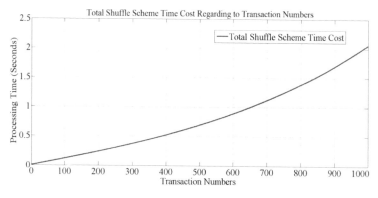

Figure 7.7: The total time cost regarding to transaction numbers.

7.5 Conclusion

In this chapter, we have proposed a novel decentralised pseudonym-management scheme for the Internet of Connected Vehicles that makes use of a blockchain based on Proof of Work in order to make the overall system more resilient to known privacy and security attacks. The proposed scheme provides a method to effectively manage pseudonyms from distribution and re-utilisation, and a pseudonym change scheme that combines physical mixed zones with virtual mixed zones. The paper discussed a number of types of vehicle privacy attacks and defence measures that are enabled by the proposed blockchain-based pseudonym management system. In addition, the proposed scheme was evaluated by network simulation tool OMNET++, including pseudonym reuse frequency, pseudonym total consumption, anonymity set size and total shuffle time. Results show that the proposed blockchain based pseudonym management system could effectively reuse existing pseudonym and pseudonym certificates, meanwhile keeping a high level of security and privacy for the ITS network system. It also cut the cost of pseudonym generation and maintenance. In addition, a total process time is computed which shows that our scheme is capable of performing pseudonym shuffling with over 1,000 blockchain transactions in 2 seconds.

References

1. Fangchun, Yang, Wang Shangguang, Li Jinglin, Liu Zhihan and Sun Qibo. 2014. An overview of internet of vehicles. *China Communications*, 11(10), 1–15.
2. Kenney, John B. 2011. Dedicated short-range communications (DSRC) standards in the United States. *Proceedings of the IEEE*, 99(7), 1162–1182.
3. Brecht, Benedikt, Dean Therriault, André Weimerskirch, William Whyte, Virendra Kumar, Thorsten Hehn and Roy Goudy. 2018. A Security Credential Management System for V2X Communications. *IEEE Transactions on Intelligent Transportation Systems*, 99, 1–22.
4. Ibid.
5. Dötzer, Florian. 2006. Privacy issues in vehicular ad hoc networks. pp.197–209. *In*: Privacy Enhancing Technologies. George Danezis and David Martin (eds.). Berlin, Heidelberg: Springer. ISBN: 978-3-540-34746-0.
6. Artail, H. and N. Abbani. January 2016. A pseudonym management system to achieve anonymity in vehicular ad hoc networks. *IEEE Transactions on Dependable and Secure Computing*, 13(1), 106–119. ISSN: 1545-5971, doi:10.1109/TDSC.2015.2480699.
7. Boualouache, Abdelwahab, Sidi-Mohammed Senouci and Samira Moussaoui. 2016. Towards an efficient pseudonym management and changing scheme for vehicular ad-hoc networks. *Global Communications Conference (GLOBECOM), 2016 IEEE*, 1–7.
8. Artail, H. and N. Abbani. January 2016. A pseudonym management system to achieve anonymity in vehicular ad hoc networks.
9. Raya, Maxim and Jean-Pierre Hubaux. 2007. Securing vehicular ad hoc networks. *Journal of Computer Security*, 15(1), 39–68.
10. Fonseca, Emanuel, Andreas Festag, Roberto Baldessari and Rui L. Aguiar. 2007. Support of anonymity in vanetsputting pseudonymity into practice. *IEEE Wireless Communications and Networking Conference* (IEEE, 2007), 3400–3405.
11. Eckhoff, David and Christoph Sommer. 2014. Driving for big data? Privacy concerns in vehicular networking. *IEEE Security & Privacy*, 12(1), 77–79.
12. Deng, Mina, Kim Wuyts, Riccardo Scandariato, Bart Preneel and Wouter Joosen. 2011. A privacy threat analysis framework: Supporting the elicitation and fulfillment of privacy requirements. *Requirements Engineering*, 16(1), 3–32.
13. Hubaux, J.P., S. Capkun and Jun Luo. May 2004. The security and privacy of smart vehicles. *IEEE Security Privacy*, 2(3), 49–55. ISSN: 1540-7993, doi:10.1109/MSP.2004.26.
14. Leman-Langlois, Stéphane. 2013. Privacy as currency: Crime, information and control in cyberspace. *Technocrime*, 135–161. Willan.
15. Sommer, Christoph and Falko Dressler. 2014. Vehicular Networking. Cambridge University Press.
16. Danezis, George, Stephen Lewis and Ross J. Anderson. 2005. How much is location privacy worth? *WEIS*, vol. 5. Citeseer.
17. Kobsa, Alfred. Springer, 2014. User acceptance of footfall analytics with aggregated and anonymized mobile phone data. *International Conference on Trust, Privacy and Security in Digital Business*, 168–179.
18. Nakamoto, Satoshi. 2008. Bitcoin: A Peer-to-Peer Electronic Cash System. Pub n place
19. Gao, Weichao, William G. Hatcher and Wei Yu. 2018. A survey of blockchain: Techniques, applications, and Challenges. *27th International Conference on Computer Communication and Networks (ICCCN)*, IEEE, 1–11.

20. Danezis, George and Sarah Meiklejohn. 2015. Centrally banked cryptocurrencies. arXiv preprint arXiv:1505.06895.

21. Zyskind, G., O. Nathan and A. Pentland. May 2015. Decentralizing privacy: Using blockchain to protect personal data. *Security and Privacy Workshops (SPW)*, IEEE, 180–184, doi:10 . 1109/ SPW. 2015.27.

22. Dorri, A., M. Steger, S.S. Kanhere and R. Jurdak. December 2017. BlockChain: A distributed solution to automotive security and privacy. *IEEE Communications Magazine*, 55(12), 119–125, ISSN: 0163-6804, doi:10.1109/MCOM.2017.1700879.

23. Lndbæk, Leif-Nissen, Daniel Janes Beutel, Michael Huth, Stephen Jackson, Laurence Kirk and Robert Steiner. 2018. Proof of kernel work: A democratic low-energy consensus for distributed access control protocols. to appear, Royal Society Open Science.

24. Dorri, A., M. Steger, S.S. Kanhere and R. Jurdak. December 2017. BlockChain: A Distributed Solution to Automotive Security and Privacy.

25. Hatcher, Gao and Yu. A Survey of Blockchain: Techniques, Applications, and Challenges.

26. Divya, M. and Nagaveni B. Biradar. 2018. IOTA – Next generation block chain. *International Journal of Engineering and Computer Science*, 7, 23823–23826.

27. Shabandri, B. and P. Maheshwari. 2019. Enhancing IoT security and privacy using distributed ledgers with IOTA and the tangle. *6th International Conference on Signal Processing and Integrated Networks (SPIN)*, 1069–1075.

28. Lei, A., Y. Cao, S. Bao, D. Li, P. Asuquo, H. Curickshank and Z. Sun. 2019. A blockchain based certificate revocation scheme for vehicular communication systems. *Future Generation Computer Systems*, 110, September 2020, 892–903.

29. Asuquo, P., H. Cruickshank, J. Morley, C.P.A. Ogah, A. Lei, W. Hathal, S. Bao and Z. Sun. 2018. Security and privacy in location-based services for vehicular and mobile communications: An overview, challenges and countermeasures. *IEEE Internet of Things Journal*, 1–1, doi:10.1109/JIOT.2018.2820039.

30. Xu, Jianliang, Xueyan Tang, Haibo Hu and Jing Du. 2010. Privacy-conscious location-based queries in mobile environments. *IEEE Transactions on Parallel and Distributed Systems*, 21(3), 313–326.

31. Zhu, Zhichao and Guohong Cao. 2013. Toward privacy preserving and collusion resistance in a location proof updating system. *IEEE Transactions on Mobile Computing*, 12(1), 51–64.

32. Chow, Chi-Yin and Mohamed F. Mokbel. Springer 2007. Enabling private continuous queries for revealed user locations. *International Symposium on Spatial and Temporal Databases*, 258–275.

33. Pan, Xiao, Jianliang Xu and Xiaofeng Meng. 2012. Protecting location privacy against location-dependent attacks in mobile services. *IEEE Transactions on Knowledge and Data Engineering*, 24(8), 1506–1519.

34. Lin, Jie, Wei Yu, Nan Zhang, Xinyu Yang and Linqiang Ge. 2018. Data integrity attacks against dynamic route guidance in transportation-based cyber-physical systems: Modeling, analysis, and defense. *IEEE Transactions on Vehicular Technology*, 67(9), 8738–8753.

35. Palanisamy, Balaji and Ling Liu. 2015. Attack-resilient mix-zones over road networks: Architecture and algorithms. *IEEE Transactions on Mobile Computing*, 14(3), 495–508.

36. Shokri, Reza, Julien Freudiger, Murtuza Jadliwala and Jean-Pierre Hubaux. 2009. A distortion-based metric for location privacy. *Proceedings of the 8th ACM Workshop on Privacy in the Electronic Society* (ACM), 21–30.

37. Kenney, J.B. 2011. Dedicated short-range communications (DSRC) standards in the United States.

38. Festag, A. December 2014. Cooperative intelligent transport systems standards in Europe. *IEEE Communications Magazine*, 52(12), 166–172, ISSN: 0163-6804, doi:10.1109/MCOM.2014.6979970.

39. Draft SAE. 2009. Dedicated Short Range Communications (DSRC) Message Set Dictionary. Society of Automotive Engineers, DSRC Committee.

40. TS ETSI. 2012. 102 941 V1. 1.1 Intelligent Transport Systems (ITS); Security; Trust and Privacy Management. *Standard, TC C-ITS*.

41. Lei, A., H. Cruickshank, Y. Cao, P. Asuquo, C.P.A. Ogah and Z. Sun. 2017. Blockchain-based dynamic key management for heterogeneous intelligent transportation systems. *IEEE Internet of Things Journal*, 99, 1–1, doi:10.1109/JIOT.2017.2740569.

42. Bao, Shihan, Waleed Hathal, Haitham Cruickshank, Zhili Sun, Phillip Asuquo and Ao Lei. 2017. A lightweight authentication and privacy-preserving scheme for VANETs using TESLA and Bloom Filters. *ICT Express*, ISSN: 2405-9595, doi:https: / / doi . org / 10 . 1016 / j . icte . 2017 . 12 . 001, http: / / www . sciencedirect . com / science / article / pii / S2405959517302333.

43. Schlegel, R., C.Y. Chow, Q. Huang and D.S. Wong. October 2015. User-defined privacy grid system for continuous location-based services. *IEEE Transactions on Mobile Computing*, 14(10), 2158–2172, ISSN: 1536-1233, doi:10.1109/ TMC.2015.2388488.

44. Lundbæk, Leif-Nissen, Andrea Callia D'Iddio and Michael Huth. Springer, 2017. Centrally governed blockchains: Optimizing security, cost, and availability. *Models, Algorithms, Logics and Tools*, 578–599.

45. Lndbæk, Leif-Nissen, Daniel Janes Beutel, Michael Huth, Stephen Jackson, Laurence Kirk and Robert Steiner. 2018. Proof of kernel work: A democratic low-energy consensus for distributed access control protocols.

46. Artail and Abbani. A pseudonym management system to achieve anonymity in vehicular ad hoc networks.

47. Nakamoto, Bitcoin: A Peer-to-Peer Electronic Cash System.

48. Merkle, Ralph C. Springer, 1987. A digital signature based on a conventional encryption function. *Advances in Cryptology, CRYPTO87*, 369–378.

49. Dwork, Cynthia and Moni Naor. 1992. Pricing via processing or combatting junk mail. *Advances in Cryptology – CRYPTO '92*, 12th Annual International Cryptology Conference, Santa Barbara, California, USA, August 16-20, 1992, Proceedings, 139–147, doi:10.1007/3-540-48071-4_10.

50. Lundbæk, D'Iddio, and Huth. Centrally governed blockchains: Optimizing security, cost, and availability.

51. Yu, Rong, Jiawen Kang, Xumin Huang, Shengli Xie, Yan Zhang and Stein Gjessing. 2016. Mixgroup: Accumulative pseudonym exchanging for location privacy enhancement in vehicular social networks. *IEEE Transactions on Dependable and Secure Computing*, 13(1), 93–105.

52. Emara, Karim. 2016. Poster: Prext: Privacy extension for veins vanet simulator. *Vehicular Networking Conference (VNC)*, IEEE, 1–2.

53. Bao, Shihan, Waleed Hathal, Haitham Cruickshank, Zhili Sun, Phillip Asuquo and Ao Lei. A lightweight authentication and privacy-preserving scheme for VANETs using TESLA and Bloom Filters.

54. Artail and Abbani. A pseudonym management system to achieve anonymity in vehicular ad hoc networks.
55. ETSI. 2012. Intelligent Transport Systems (ITS); Security; Trust and Privacy Management. Technical Report, ETSI TR 102 893, European Telecommunications Standards fffdfffdfffd.
56. Tomandl, A., F. Scheuer and H. Federrath. (October 2012). Simulation-based evaluation of techniques for privacy protection in VANETs. *IEEE 8th International Conference on Wireless and Mobile Computing, Networking and Communications (WiMob)*, 165–172, doi:10.1109/WiMOB.2012.6379070.

Simulation Platforms for Autonomous Driving and Smart Mobility: Simulation Platforms, Concepts, Software, APIs

Lars Creutz[1]*, Sam Kopp[1], Jens Schneider[1], Matthias Dziubany[1], Yannick Becker[2] and Guido Dartmann[1]

[1] Institute for Software Systems (ISS), Trier University of Applied Sciences, Birkenfeld, Germany
[2] Trier University of Applied Sciences, Birkenfeld, Germany

8.1 Introduction

In the near future, smart mobility will lead to major changes in the way traffic works and affects its participants and its environment. Not only will autonomous vehicles change how we use and experience mobility itself, but also, emerging services are about to create entirely new user experiences when interacting with the environment and the surrounding traffic. To reduce risks and costs of the design of algorithms for autonomous vehicles, simulation tools are developed which allow algorithm developers to test multiple scenarios and machine learning algorithms virtually, before testing it in physical environments with real cars. This chapter gives a brief overview of existing mobility simulators and their architecture, compares open source functionality with commercial solutions and describes, how application interfaces are designed, to communicate with the simulation systems. We also discuss, how these simulators can be used besides artificial intelligence – to develop new smart services. Similar to the publications [37, 52, 54], we distinguish three general classes of traffic simulation models.

Macroscopic approaches basically focus on the collective flow of entities within a system. The individual behaviour of a vehicle within this flow is not

---

*Corresponding author: l.creutz@umwelt-campus.de

Table 8.1: Overview of traffic flow models according to the
categories of [52] and [37]

Model	Features
Microscopic	Description of the dynamics of an independent vehicle
Macroscopic	Based on vehicle density and traffic flow
Mesoscopic	Mixture of micro- and macroscopic models

of interest. Traffic is simulated from a high-level point of view, as a stream of numerous vehicles [37]. It is used to observe and analyze the flow of traffic, regarding vehicle density and distribution, velocity or lane usage in larger road networks [45].

Microscopic simulations, however, deal with the behavior of a single vehicle within a traffic system. This includes the actions and reactions a car performs, to move in a given environment, with respect to traffic constraints and other agents [37, 44].

As stated in [37], mesoscopic models are used to find a compromise between the macroscopic level of simulation and the simulation of individual interactions between vehicles.

The following chapter gives an overview of existing mobility simulation tools, not solely dealing with traffic simulators, but also presenting ADTF, a framework for developing advanced driver assistance systems, up to completely autonomous driving systems [71]. The tools are described in terms of their fields of application, general functionality and features, communication and data formats, as well as outputs, to evaluate simulation results.

8.2 Eclipse SUMO

Eclipse SUMO is a portable road traffic simulation system, written in C++ and designed to handle road networks in a large scale. The initial release of the system was in 2001 by the Institute of Transportation Systems at the German Aerospace Center and it is still under consistent development [51]. The fact that the application is released under the Eclipse Public License V2, causes the project to be open source and publicly available [72]. Its name is derived from "Simulation of Urban Mobility".

8.2.1 Field of application

Before we go into more depth, concerning the overall structure and concepts of Eclipse SUMO, we like to begin with a brief overview of research topics the system was used in. First of all, the ability to model traffic lights was used to evaluate their performance in combination with new algorithms, to improve timing and planning [49]. A second field, where Eclipse SUMO is heavily used, is in optimizing routes and the choice made while navigating, for example by increasing the performance of dynamic user assignment [35]. One of the most important research topics is the field of vehicular communication, especially vehicle-to-everything (V2X), where the simulation is coupled with a network simulation [50, 65]. The possibility to model entire networks (for example with ns-3 [25] in combination with SUMO's Traffic Control Interface) allows us to test and develop new services, that interact with our environment, based on microscopic simulation data.

8.2.2 Functionality

In [54] multiple functions are already explained. We will shortly summarize the most important functions to give an overview for interested readers. Eclipse SUMO can be classified as a microscopic simulation platform which includes pedestrians, vehicles, public transport systems and time schedules for traffic lights, inside the environment. The platform consists of various applications, which can be used to conveniently interact with the simulation itself. SUMO describes the command line interface for the entire simulation. The graphical user interface is encapsulated and named SUMO-GUI. Besides the actual simulator, three important tools generate networks, to run the simulations on: NET-CONVERT is a tool, which allows the user to import road networks and scenarios in various formats [54] (see 8.2.3), NETGEN generates abstract networks and NETEDIT is a network editor, that uses a graphical user interface to edit, add and delete the network elements [54], for example, the streets themselves or traffic lights. Besides the generation/editing of networks, Eclipse SUMO is shipped with several routing components. The tool JTRROUTER is used to generate routes, which are based on turning ratios at intersections. DUAROUTER is a component, which generates routes, based on dynamic user assignation. In contrast to the previously mentioned tool, MAROUTER describes the process of macroscopic user assignment, derived from capacity functionality [54]. Lastly, DFROUTER can be used to generate routes via detector data [54]. Besides the mentioned tools, the most common way to interact with SUMO is through its "Traffic Control Interface" (TraCI). It allows the user to take control of the entire simulation and interact with the simulated traffic online. Implementation-wise, TraCI is a TCP based client/server architecture, whereas SUMO acts as a server which listens on a specific port for control instructions (see 8.2.3 on how to interact with SUMO by TraCI).

8.2.3 Communication and Data Format

The simulation interface indirectly accepts various data formats as network input by using NETCONVERT, to generate native SUMO xml-based network descriptions. SUMO supports several important data formats, to be able to interact with other simulation engines. The most common convertible formats are described in [54] and illustrated in terms of import/export capability, which is not bidirectional in Table 8.2. Also, NETCONVERT is able to handle other formats, which were developed internally by the German Aerospace Center, like an edited specification of the Navteq Geographic Data Files [24].

Table 8.2: Overview of formats used with NETCONVERT

Format	Import	Export
Native SUMO	Yes	Yes
SUMO	Yes	Yes
OpenStreetMap	Yes	No
VISUM	Yes	No
VISSIM	Yes	No
OpenDRIVE	Yes	Yes
MATsim	Yes	Yes
Shapefiles	Yes	No
Robocup Rescue League	Yes	No

In general, NETCONVERT, used on non SUMO-formats, takes the to-be-converted network format and generates the corresponding SUMO file. Besides road networks, further data formats are supported to import public transport stops (from OpenStreetMap data), public transport lines, street signs, street names, parking areas and railway information. In addition to the conversion of several data formats for the simulation to use, NETCONVERT provides functionality to fix missing network data, to ensure the necessary level of detail for a microscopic simulation [54]. The general SUMO road networks can be seen as graphs, where intersections are represented as nodes, connecting the unidirectional edges (roads), which, in turn, consist of a number of lanes. Several rules also apply to the described geometry. Therefore, a lane holds information about the allowed classes of vehicles on it and additional information, like speed limits. Intersections include information about traffic lights and right-of-way rules at the specific area. All this additional information is not only a key factor on a realistic approach, when using Artificial Intelligence on Eclipse SUMOs models, but also allows the user to quickly get reliable information about a scenario without the need of manually labeling data beforehand. An entire scenario is created by combining the mentioned graphs with additional information. The demand/activities, contained in a scenario, are defined as single routes/flows, where pedestrians can be included, using different interacting models. The first model is the non-interacting one, where pedestrians are just included as individuals, which do not interact with vehicles or public transport systems. Refer to Table 8.3, to examine the fields of usage per model. Eclipse SUMO defaults in using the striping model, where pedestrians in fact interact with the traffic and explicitly occupy space inside the simulated environment. Furthermore, there are two models defining how vehicles are allowed to overtake. The sublane model, which describes the possibility to overtake on a single road, if there is enough space and the oncoming lane model, which allows to use the lane of the oncoming traffic for overtaking [50].

Table 8.3: Overview of interaction models

Model	Example usage
Non-interacting	Accumulation of people with delayed public transport
Striping	Simulation of real-world traffic with pedestrian interactions
Oncoming lane	Simulation of real-world driving behavior
Sublane	Heterogeneous traffic simulation

After the brief description of a scenario setup, we now describe the common way of interacting with the simulation, by using the Traffic Control Interface (TraCI) [54]. The system itself is a TCP client/server architecture, whereas the core simulation acts as server and waits for incoming connections [28]. Using the TraCI results in taking control over the entire simulation system, which causes

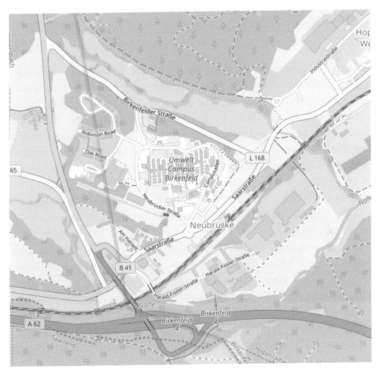

Figure 8.1: Screenshot of Eclipse SUMO tool "osmWizard" – Section of Open StreetMap which will be converted to a SUMO network file.

Figure 8.2: Screenshot of Eclipse SUMO-GUI – Imported map inside the simulation.

every simulation step to be triggered by the client. The actual simulation core (SUMO) just does initial work by setting up the environment, before waiting for incoming connections to take over. The currently used protocol, described in the official German Aerospace Wiki for Eclipse SUMO, defines the TCP messages as containers for control instructions or results. The message itself consists of a header, the length of the entire message and the commands. A command is described by an identifier, general length and its content. The container's width is fixed 16 bit. We use the terminology block to describe a part of the message, which is exactly 16 bits. The first 32 bits (2 blocks) describe the header and the length of the total message. The following data describes the commands and is divided into two blocks each. The first block contains the length with a fixed size of 8 bits and an unambiguous identifier, whereas the second block holds content to the command/result. If the command's length exceeds 255, the protocol defines an alternative message format, which can be found in the official specification [28]. The unique message identifiers are described on the mentioned wiki and shall not be in the scope of this chapter. We recommend using given APIs to interact with the simulation's core. The best coverage is given by the python package traci, which supports all TraCI commands. Several other programming languages are supported, too, but the coverage varies [27].

8.2.4 Output

The Eclipse SUMO suite offers various outputs while simulating traffic. First of all, the actual visible output of the graphical user interface (SUMO-GUI) needs to be mentioned. The user is able to review the entire, defined scenario and, correspondingly, the behavior of vehicles, pedestrians and the infrastructure (e.g. traffic lights). Nonetheless, it can be quite hard to review a huge network, therefore, SUMO provides several output formats, which can be classified in the type of output: file, socket or stream. Information about vehicles can be obtained in aggregated and disaggregated form, to gather almost any information about the vehicles: trajectories, speeds, their trips, routes (with detailed information about stops, which are used to simulate the drop-off of persons or containers), their battery usage (on electric vehicles) and their emissions [26]. Furthermore, information about simulated detectors can be gathered, which, for example, include the usage of inductive loops. Additional information about traffic lights, network elements (lanes/edges) and junctions can also be part of the simulation's output. In general, all those output files provide a central way to incorporate data from SUMO to other platforms (by providing generators/converters) and evaluate and measure the quality of the simulation.

8.2.5 Conclusion

Eclipse SUMO is especially useful when simulating on a microscopic level, with the ability to interact with the simulation through the stated API. This allows the simple application of Artificial Intelligence techniques, which can be easily integrated into the simulation or trained from the simulated data and change the environment. This could be used to learn new driver or traffic models. Furthermore, the graphical output is not very challenging, which allows SUMO to be used on non-state-of-the-art hardware, allowing users to participate in this specific field on a lower budget. Eclipse SUMO is great to develop and test new smart services. Consider, a future smart city with digitally monitored environmental zones, for example. Due to the possibility to simulate any emission data inside SUMO, we can define new smart services in the form of contracts between the vehicles and the environment. Those contracts can for instance be connected to modelled inductive loop detectors or cameras, to perceive the entrance and exit of non electric vehicles (by the properties inside the simulation) and calculate an emission fee for the time period they spent inside a specific area.

8.3 MATSim

MATSim is an agent-based, microscopic traffic flow simulation system implemented in Java, that originates in Kai Nagel's work at the Institute for Transport Planning and Systems at the Swiss Federal Institute of Technology Zurich [47]. The project is open source, which makes the application not only accessible to everyone, but also extendable. The structure of the project consists of multiple modules, which can be replaced, used stand-alone or used as a whole. Its name is derived from Multi-Agent Transport Simulation.

8.3.1 Field of Application

MATSim was successfully used in various projects all around the globe, some of which made their scenario data publicly available, including networks and populations [23]. In general, the usage included public, private and freight transport. Some projects contained walking, biking and taxi traffic, too. Besides the high variety of simulating different traffic, the projects also used different input formats and data sources, which shows the high support interoperability between MATSim and other systems in similar domains. A brief, consistently-updated overview of projects can be found on the official MATSim homepage [20] or partly in [60], [48] describes several detailed scenarios in chapters 52-96. Bischoff and Maciejewski [36] used MATSim to show, how a fleet of electric taxis can be used on small cities, without any negative impact on practicability for passengers, with a relatively low amount of charging stations. This was possible due to the features of MATSim, to model an electric vehicle in a very detailed way, including a model of stationary charging, which was included in the simulation.

8.3.2 Functionality

MATSim supports large-scale scenarios and allows the user to simulate public and private transport. Also, MATSim currently offers frame-works for demand-modeling, agent-based traffic simulation, re-planning and analyzing tools [20]. Before we describe the workflow, we need to introduce the so-called "MATSim loop" [47], which illustrate the simulation processes.

The available modules plug in at different points of the loop or act on a global scale. Due to the constant development and new features, we refer to their documentation [22]. Besides the included modules, it is also possible to extend the open source project to fit your needs and integrate new features. [70] describes the contribution process and detailed guidelines for the development of custom extensions. Figure 8.4 shows the available modules at their respective position in the loop. The common way of how to configure a module, is described in 8.3.3.

Figure 8.3: MATSim loop/cycle. Figure similar to Horni et al. [47].

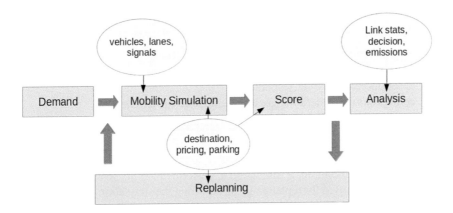

Figure 8.4: Some MATSim modules within MATSim loop. Figure similar to Horni et al. [47].

As presented in [47], MATSim was designed to simulate a single day, using a co-evolutionary algorithm to optimize each agent's daily schedule, where agents compete on space-time slots.

In [59], the procedure to setup a simulation is presented. At the beginning, the scenario has to be modelled. This is done by creating or providing xml files,

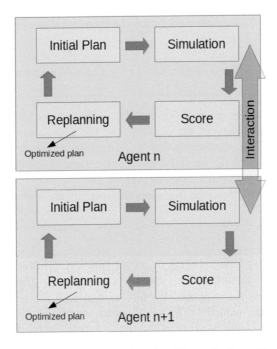

Figure 8.5: MATSim Co-Evolutionary algorithm. Figure similar to Horni et al. [47].

that describe its different components. For a simulation of public transport, for example, the following components would have to be modelled:

- graph, defining the routes available for public transport
- description of the available vehicles
- description of the time schedule
- configuration file defining general settings for the simulation.

These can be for example the start and end time of the simulation, the agent's behavior or the method of rescheduling.

Afterwards, the loop is started, based on the previously given specifications. During the loop execution, a simulation is performed and its score is calculated. This score is then used for replanning, i.e. adjusting the simulation. Then, the loop starts again from the beginning.

8.3.3 Communication and Data Format

The application heavily relies on xml-based configuration files, which are used to setup the entire suite [59]. Every module can be configured in a single configuration file, which itself is based on an xml-format. To clarify, the following excerpt describes the configuration of the network and plans module and was taken from the official MATSim example project [21].

```
<module name ="n e t w o r k">
  <param name ="i n p u t N e t w o r k F i l e" value ="n e t w o r k . xml" />
</module>
<module name ="p l a n s">
  <param name ="i n p u t P l a n s F i l e" value ="p l a n s 1 0 0 . xml" />
</module>
```

The listing above gives a first insight, on how a scenario is defined, by specifying the definition of the network and the agent's plans, among other parameters [59]. The minimum possible implementation sets up a configuration, creates a scenario with the generated Config object and passes the Scenario to the Controller, which executes the simulation [22].

Horni et al. [46] describe five ways of accessing the simulation, which are stated in Table 8.4.

8.3.4 Output

MATSim generates a high variety of output data and writes it to multiple files while running the simulation. This data is normally used to analyze the obtained results and scores of the agents. We like to start with the generated iteration folders, which are named "ITERS/it.{iteration number}" and located inside the output folder (same location as the configuration file by default). For each iteration, the entire simulation process is monitored, while running. Before we

Table 8.4: Levels of access to MATSim

Access	Description
GUI	Usage of graphical user interface with modules network, config and population
GUI with additional configuration files	Mostly used to simulate public transport and car traffic
Java Scripts	MATSim supports Java as scripting language, to setup the simulation. The usage has been described in the previous paragraph
Extensions	Usage of external contributions (might have no connection to official MATSim project)
Own extensions	Use your own extensions, which can also be contributed to the project

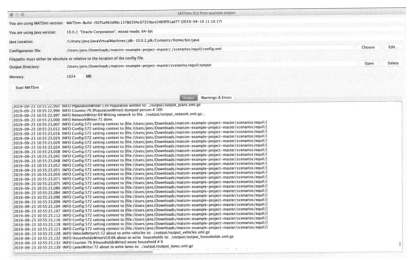

Figure 8.6: MATSim Graphical User Interface.

describe, how to visualize the output data, Table 8.5 lists some of the available output information [61]. This output can be visualized by using a variety of tools. Therefore, we would like to introduce Via [29]. The application, developed by Simunto, is able to visualize most of MATSim's file types. Figure 8.7 shows the official MATSim example network, imported in Via. The generated data can be used for analyzing purposes, for example, to determine, what happened in general during the iterations (events). How did the agent's plans change over time (plans), e.g. choosing a different route after a period of days/learning? Which days are most time consuming for an agent to get to work (trip durations) or the traffic volume in general on a given point of interest (leg histogram)?

Table 8.5: MATSim output data, see [61]

Output type	Description
Events	Record of any activity (e.g. a person entering a vehicle) with additional information about the agent, generated after every 10th iteration per default.
Plans	Current state of the agent's plans after a set number of iterations (except the final plans, which are located in the top-level of the output directory)
Trip durations	Output of number of trips and their duration for each pair of activities (e.g. home to work)
Leg histogram	Histogram about arriving and leaving agents per time unit

8.3.5 Conclusion

MATSim is suitable for large-scale scenarios, containing public and private traffic models with a huge amount of detail to the simulation. Its output format allows for further analysis of important events during the simulation, for example, how agents behave differently over a period of iterations. The open source nature of the project concedes it to be used in any kind of domain, including commercial products/organisations without the need of change inside the users infrastructure. Another benefit of using MATsim is the ability to customize the system by modifying core components or extensions. Those kind of features allow MATSim to be used in very specific fields of Artificial Intelligence by putting entirely new sets of information in context, which influence the target function of an algorithm or neural net.

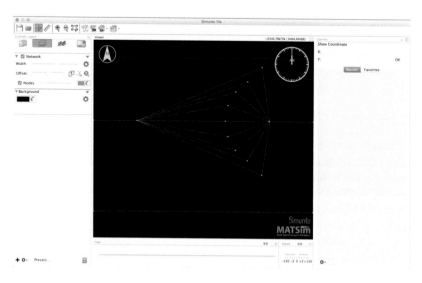

Figure 8.7: Simunto Via used to visualize MATSim network data.

8.4 Apollo Dreamland

Apollo Dreamland is a simulation engine, developed by Baidu, and a part of the open Apollo.Auto autonomous driving platform [2]. Apollo.Auto can be seen as a microscopic simulator, based on an open-source software stack, which provides all necessary modules for running an autonomous vehicles. Written in C++, former versions of Apollo.Auto were relying on ROS (Robot Operating System), but the most recent versions use the new self-developed messaging framework CyberRT [4]. The software is divided into a multitude of platforms, to cover all necessary areas for development [3]:

(1) Cloud Service Platform: Simulation, HD Maps, etc.
(2) Open Software Platform: Vehicle controlling, routing, planning, etc.
(3) Hardware Development Platform: Vehicle hardware and sensors, e.g. CPU, GPS, Cameras, Lidar, etc.
(4) Open Vehicle Certificate Platform: Certified Apollo Compatible Drive-by-wire Vehicle, Open Vehicle Interface Standard

Apollo Dreamland is the component, that takes care of vehicle simulation in a microscopic manner. It is hosted as a web-based cloud service for testing and developing driving algorithms, compatible with Apollo's autonomous driving framework [8]. The tool features a 3D visualization and a wide variety of customizable options such as different execution modes, a large number of automated simulation metrics and many executable simulation scenarios. These scenarios can either be generated manually, be more deterministic with well-defined behavior or derived from real-world data [8]. The platform is currently in development and most of the functionality is only available as private beta test [2 , 10].

8.4.1 Field of Application

Since the Apollo platform is a rich ecosystem, that offers functions for all tasks concerning autonomous driving, it is versatilely used in research and industry projects [2, 41, 42] and rarely reduced solely to its simulation component Dreamland. The latter is mainly employed for supporting the development, testing and evaluation of autonomous driving software and algorithms, which is particularly considerable, when the development is also conducted on the Apollo platform. For example, Xu et al. followed this approach in [69]. First, they created an automatic, learning-based procedure to extract a dynamic vehicle model from real-world driving data, to represent different car dynamics (e.g. brand, age, degeneration of parts) in simulation. After that, they integrated their method into Apollo Dreamland to test and evaluate its efficiency. Deviating from this, Chao et al. introduced a method to improve interactions between vehicles and other road users (bicycles, pedestrians) in simulation environments, using a force-based concept [38]. They considered Dreamland a state-of-the-art simulator and therefore compared their results with Dreamland's built-in interaction dynamics, in order to evaluate their development. Similarly, Fang et al. presented a framework, able to simulate 3D Lidar point clouds, that can be used for training autonomous vehicles, and also compared their results with the generated Lidar data of Apollo's simulator [43].

8.4.2 Functionality

Dreamland's goal is to make fast and reliable testing of autonomous driving algorithms available for developers. A good overview of the general functions is given in [8, 9]. As illustrated there, the tool is able to execute the implemented algorithms in a complex simulation environment, that offers features like a complete road infrastructure with intersections, traffic lights and several types of lanes and junctions. Furthermore, there are various obstacles, such as bikes, pedestrians, etc., and different driving plans to choose for the controlled vehicle. Simulations can also be accelerated, to run more loops in a shorter amount of time. Since Dreamland is provided as a cloud service, the tool promises a large capacity of computing power for this intended use. Generally spoken, the web-based architecture makes it possible to use the simulation engine without the need to install large, complex software or purchase expensive high-performance hardware. This is a big difference to similar tools in this chapter.

Two scenario types are supported by the platform [8]: *worldsim* and *logsim*. Worldsim scenarios are well-defined, which means they offer a predetermined behavior for road-users, obstacles or traffic lights. In return, Logsim scenarios use data, captured in real-world conditions. In most cases, this data is more complex and less deterministic, but closer to reality, as agent behavior can be unpredictable and fuzzy, so traffic conditions become more complicated.

About 200 pre-defined scenarios are provided by the platform and additional scenarios can be created by the user [8].

While simulating the individual test cases, the tool uses an automatic grading system with a total of 12 metrics, to give an appropriate feedback to the developer. Each test run is graded in various categories, e.g. vehicle parameters, collisions, red-light violations or speeding [8].

8.4.3 Communication and Data Format

As indicated before, the software is only available as an online service on the Apollo website [2]. Scenarios can be added via an interactive editor in the browser, on pre-defined maps. As described in the official documentation [12], each scenario has to include the simulated car with its initial state (position, heading, speed, acceleration) and basic, descriptive information and metrics, needed to determine the success. In order to make the scenario meaningful, different types of obstacles can be inserted. These can be cars or pedestrians, that are static or mobile. A movement can be triggered by time or by distance to the controlled vehicle. The created scenarios can be executed along with the provided sample scenarios in groups. To test algorithms created by the user, they must support the Apollo modules standard API. The procedure is described in [11] as follows:

The API is defined as ROS or CyberRT topics in the form of protobuf interface files. The recommended way to use Dreamland for simulation is to fork the Apollo git repository first. In this version, any changes can be made to the testable modules. Afterwards, the forked Apollo version will be provided to Dreamland for simulation. Currently, only public git repositories can be used for this purpose. Every piece of software to be tested has to communicate via the Apollo API. This aspect, unfortunately, narrows the field of application to only apollo-compatible implementations of autonomous driving algorithms [10].

During testing, Dreamland supplies data for four modules, according to Table 8.6. As stated in [11], the planning module, in Dreamland's current state, is always subject to test and must therefore be provided by the user. The output of the prediction module can be included in the simulation, but is optional. For simulation, the software to be tested must be available as a public git repository. The content of this repository is then automatically built for testing and used in the simulation.

Table 8.6: Apollo module output provided by Dreamland [11]

Module	Information
Localization	Position, heading, orientation, velocity and acceleration of the autonomous vehicle
Perception	ID, size, position, heading, type and polygon points of perceived obstacles; ID and color of detected traffic lights
Canbus	Vehicle speed in meters per second
Routing	The routing response, consisting of road segments to use, in order to reach the destination

8.4.4 Output

The result of a simulated test run consists of multiple parts. At first, an overview over succeeded and failed test runs is given with the graded metrics, we mentioned before [11]. In a more detailed view, every test case can be inspected for success or failure in the various categories. Also, it is possible to view a 3D visualization. This contains an illustration of the scene with road conditions, lanes, traffic lights, obstacles, routing, planned trajectories and more in real-time. Also, the vehicle status is shown with parameters like velocity, braking and so forth [9]. In addition, individual log/bag files for each test case are available to download [11]. These files contain the detailed message flow between all system components and can be used for replay and debugging purposes.

8.4.5 Conclusion

Apollo Dreamland is a powerful simulation tool, that is able to support the development of autonomous driving software as a whole. Although using the Apollo API is mandatory to communicate with the tool, this restriction can be beneficial, if the rest of the development is conducted on the Apollo platform, too. Since Apollo offers an entire software stack for autonomous driving, it is possible to set up a development environment, that offers all necessary modules with a high compatibility between them.

Another advantage over similar tools is the web-based architecture, that requires no extraordinary hardware. Additionally, Dreamland provides a big amount of pre-defined scenarios and a comprehensive grading system out-of-the-box.

Nevertheless, the platform is still in development and currently, only the versions of the planning and prediction modules, provided by the user, can be tested, which represents a limitation of the possible uses [10, 11].

8.5 LGSVL Simulator

The LGSVL Simulator, developed by LG Electronics, offers a simulation environment for testing and developing autonomous vehicles on a microscopic level [15]. As stated in [19], LGSVL Simulator aims to make simulations of a whole vehicle as realistic as possible, at low cost and in short time. Therefore, the tool provides a 3D simulation with HD-rendering [15]. The product can be used for testing and validating complete autonomous-driving software stacks, working as a whole, including an integration with TierIV's Autoware, Baidu's Apollo Platform as well as ROS [19]. The simulator is developed with C# and based on the Unity 3D [73] game engine. It is made publicly available under a custom license, including the source code. With the first release candidate available at the end of 2018, it is a rather young product, which is still under heavy development. The following is subject to the latest release at the time of writing, which is version 2019.09 [16 , 17].

8.5.1 Field of Application

Since LGSVL Simulator is a relatively new product, it is hard to find projects, the tool was already used in. However, it is definitely considered as a potent simulation environment in recent literature such as [68], [67] or [62]. One of the few applications is described in [53]. There, the authors introduce a method to fuse multiple cameras on an autonomous vehicle, in order to improve object detection. They are developing on the Apollo platform and use the LGSVL simulator to test their algorithms. Another project by Piazzoni et al. [58] dealt with the evaluation of perception and decision-making subsystems in an autonomous vehicle. They also used Apollo to implement their system proposal and LGSVL simulator as testing environment, to be able to validate their researched model.

8.5.2 Functionality

A summary of the tool's functionality can be drawn from the official documentation [15]. As stated there, the simulator provides a variety of virtual worlds and vehicles. First of all, the environment consists of vehicles, roads, crossings, walkways, pedestrians, traffic lights and buildings of different kinds. Non-playable, automated road users are supported out-of-the-box. This enables a full simulation of inner-city vehicle traffic, that behaves according to adjustable traffic rules. These rules are, for example, using the correct lanes, preventing collisions, stopping at signs or red traffic lights, and are stored in the map file. This type of simulated traffic is usually not deterministic, but the simulation environment supports the use of a fixed seed for a particular scenario, in order to ensure that its sequence remains identical. The simulation of pedestrians is possible in the same fashion.

The autonomous vehicles include a variety of integrated sensors, e.g. Lidar, radar, GPS, IMU and cameras. The sensor data is made available and can be used by autonomous software stacks, to control the car in the virtual environment. To add more variety to the scenarios, it is possible to change the weather conditions and the time of day, even while the simulation is running. The possible weather conditions are rain, fog, cloudiness and street wetness. Furthermore, the open architecture of the software allows to integrate self-made vehicle models and maps via the Unity 3D editor.

A single simulation scenario can be run infinitely or for a specified time, to allow step-by-step simulations. It is also possible to simulate faster than real-time, if the performance of the used hardware is sufficient. In general, the tool is very compute-intensive, since it provides a highly-realistic simulation environment with full HD support and real-time rendering of sensors, like Lidar as well as several camera perspectives. Therefore, a powerful graphics card with at least 8GB memory and a 4GHz quad-core CPU is required for using this product.

8.5.3 Communication and Data Format

The LGSVL Simulator features built-in compatibility with two open source software stacks for autonomous driving: Apollo.Auto and Autoware.AI. It also includes communication layers for ROS and ROS2 [19]. To interact with the simulation and change characteristics, the tool provides a Python API with a wide range of features, from choosing the map, changing weather conditions, pedestrian movement and traffic lights, to controlling vehicles other than the main car [15]. In addition to that, it is possible to create specific simulation scenarios via the API. For example, actions or dynamic events, triggered by the time or position of the simulated vehicle, other vehicles or pedestrians, can be set up [18]. The interface for the Python API is defined in a configuration file, that has to be stored in the root directory and describes how the API is supposed to be used, e.g. through a local port or remotely via network [18].

The documentation [15] depicts, how new vehicles and maps can be created in the Unity 3D project. It provides tools to annotate lanes and traffic lights. Afterwards, they can be exported as high definition maps into the Open DRIVE format or the format for usage with Apollo.Auto or Autoware.AI, respectively. High definition maps, including information about lanes, intersections, traffic light positions and speed limits are needed by autonomous-driving software for safe operation. Pre-made vehicles and maps can be imported into the LGSVL simulator with an integrated web interface.

8.5.4 Output

One of the main features of the LGSVL Simulator is its high-definition visual display [19]. The visualization enables the user, to watch the simulated scenarios in real-time, featuring multiple camera perspectives (see Figure 8.8). Additional renderings can be activated. This, for instance, includes a visual representation of the Lidar point cloud. The Python API offers possibilities to log custom output of automated simulation runs, e.g. with callback functions on collisions, holding information of time, position and involved vehicles [18]. By using the API to

Figure 8.8: Three camera perspectives of LGSVL Simulator [34].

properly log and execute test runs, it is possible to achieve a statistically verifiable number of results in a manageable amount of time [34].

8.5.5 Conclusion

LGSVL Simulator provides a sophisticated simulation environment for testing and developing single driving algorithms up to entire software stacks. It features a highly-realistic rendering, but requires high-performance hardware in return. An advantage over similar tools is the compatibility with Apollo and Autoware platforms as well as ROS. Due to its architecture, the simulator can be run on a developer's computer in real-time, to run tests, directly integrated into any kind of development process. For further and more extensive runs, the simulation can be spread across multiple systems, including cloud-computing services. Since ground truth data is available during simulation, the system can also be used to train machine learning algorithms, e.g. for object detection and collision prevention.

8.6 CARLA

CARLA (Car Learning to Act) is an open-source simulator for urban driving environments and can be considered a microscopic simulation platform. It has been developed by the Computer Vision Center at the Autonomous University of Barcelona and was created to support development, training and validation of autonomous driving systems for single vehicles [39]. The platform offers a wide variety of open assets, such as buildings, vehicles, pedestrian models and urban layouts [39]. Furthermore, there is a highly configurable simulation environment, that provides a flexible specification of sensors, traffic settings, weather conditions and even map generation [7]. Below, the latest stable release is reviewed, which was version 0.8.2 at the time, this chapter was written.

8.6.1 Field of Application

The main purpose behind CARLA is to develop and test autonomous driving software. For example, Dosovitskiy et al. [39] evaluated different approaches to autonomous driving and their performance, namely a classic modular pipeline, imitation learning and reinforcement learning. They used the CARLA simulator to train and compare the three methods in various scenarios, differing in environmental conditions and difficulty. Osinski et al. [57] followed a similar procedure, but concentrated solely on developing a reinforcement-learning method. In the latter study, the CARLA simulation was chosen to train the neural network and create a driving policy, which was later deployed to a real-world vehicle.

8.6.2 Functionality

The system is written in C++ and based on the Unreal Engine 4 [74] as its core software. The engine takes care of rendering a realistic, graphical 3D environment with static objects like buildings, vegetation, roads and traffic signs, as well as dynamic objects like vehicles and pedestrians. In addition, it handles real-world physics and a basic intelligence for non-player agents. So, vehicles and pedestrians follow a fixed logic to move around realistically, e.g. by respecting junctions, traffic signs, speed limits or avoiding collisions [39]. Finally, the Unreal Engine also functions as an unified ecosystem for the simulation. Thus, the system can easily be extended with plugins and assets.

The simulation offers a wide variety of illumination and weather scenarios to train autonomous driving in different conditions. Also, the user can choose the sensors, the car should be equipped with, to help mapping a real-world autonomous vehicle to the software. CARLA is able to simulate a GPS sensor, a Lidar and multiple cameras, including a depth camera [6, 39]. The corresponding outputs are explained in 8.6.4.

CARLA can be executed with or without a graphical visualization and therefore offers to select two different time modes. Users may select a fixed time step and specify the desired frames per second, to run the simulation as fast as needed or set a variable time step, which keeps up with real time and adjusts the frame rate slightly on each update [6]. The first option is especially useful, when a learning algorithm has to be trained. For this matter, the simulation can be looped and run as quickly as possible, to train one or more scenarios.

When the simulation is performed in a graphical environment, CARLA is heavily relying on a good GPU. Hence, the hardware is recommended to have at least enough performance to run the Unreal Engine.

8.6.3 Communication and Data Format

As described in [39] and [6], CARLA has been implemented as a server-client system, that communicates data via sockets. The server is responsible for running the simulation and rendering the scene. To control the simulation, there is a client API, programmed in Python, that opens a socket and listens for instructions. The client can be any external application which sends commands to control the vehicle, including steering, acceleration and braking or meta-commands to alter the simulator settings. For the latter, the possibilities range from resetting the simulation, respectively the loop, to modifying attributes like the vehicle's starting location, vehicle sensors, density of cars and pedestrians, weather, illumination and more [39]. It should be noted that, due to this architecture, CARLA can be run as a distributed system, where a high-performance server runs the simulation and the clients control the execution. CARLA also supports multiple clients, which in turn are able to control multiple agents in the scene. Each client receives sensor data from the simulated vehicle in return.

The system can be deployed in two different modes [6]:

(1) *Server mode:* The simulation is run as described above. The server waits for a client to connect to the socket and send commands. Afterwards, it reads the

given data, executes the instructions and sends a response. The simulation can thus be controlled programmatically.

(2) *Standalone mode:* The server-client logic is hidden in the background and the simulation is behaving like a video-game. The user can control the vehicle with keyboard inputs, that are processed by the Unreal Engine, which also acts as a client application. The output and parameters are visualized for the user.

Simulation settings and environmental parameters, for example traffic and pedestrian density or weather, are initially read from a file, before a scenario is started. However, the file can be modified prior to execution or, if the server mode is used, these properties may also be changed at run time.

8.6.4 Output

As already mentioned, all clients receive a response with simulation data from the server. Every frame, a package with measurements and image data is sent out via the API and can then be processed by the client. Table 8.7 summarizes the output data, CARLA generates, as extracted from the official documentation.

Table 8.7: CARLA output data [6]

Output	Description
Timestamps	Every frame is described by a frame number, the timestamp of the current frame and the timestamp of the elapsed time, since the simulation started.
Vehicle transform	World location, orientation and rotation of the agent, as well as acceleration and speed.
Collision variables	Describes a collision intensity, split in three distinct figures: collision with vehicles, collision with pedestrians and collision with other objects.
Lane/Off-road intersection	Percentage of the vehicle, intersecting with the opposite lane or the outside of the road, given as two separate values.
Autopilot control	Contains a suggestion for control values, that the in-game autopilot system would apply to this frame. This includes values for steering, throttle, brake, hand brake or reverse gear.
Non-player agents info	This output is optional and has to be particularly enabled. It delivers information about other agents in the scene. These are vehicles, pedestrians, traffic lights and speed-limit signs. The list contains data to describe their movement (transform, speed, etc.) or their state and value, in case of traffic lights and speed limits.

Further output is gathered from the sensors and cameras, the virtual vehicle is equipped with. CARLA has the ability to generate images from the scene and send it back to the client, where the pictures can be processed and analyzed by the tested algorithms, which is similar to a real-world scenario. There are five different post-processing effects to render an image, in order to simulate the four different cameras and a Lidar sensor on the car. The available cameras are described in Table 8.8. All images are encoded by the server as a BGRA array of bytes [6]. It is up to the client to parse these images in a desired manner.

In addition to the cameras, there is also a virtual Lidar sensor, implemented with ray-casting. The data is sent every frame and contains information about the angle the Lidar is working in, the number of channels, the amount of points per channel and all captured points as a point cloud [6].

8.6.5 Conclusion

CARLA can be especially useful for developing driving algorithms for autonomous vehicles. The system allows testing and training programs in an environment, comparable to the real world, as it uses realistic constraints and scenarios. It also provides similar inputs and sensors to those of a real autonomous vehicle, in order to make the created algorithms more adaptable. Especially machine learning solutions can efficiently be trained and enhanced, because of the constant feedback and the fast simulation loops.

Table 8.8: CARLA cameras [6]

Camera mode	Description
Scene final	A simple view of the scene is rendered and modified with some post-processing effects, like vignettes, blurs, etc. This mode aims to generate a realistic picture, to imitate a real world camera.
Depth map	The virtual depth camera generates a map with 24 bit floating precision, point codified in RGB. The output is a gray scale image for display, but the corresponding point cloud can be extracted too.
Semantic segmentation	In this picture, each object within the view is classified by a different color, according to the object class.
None	A regular image of the scene with no special processing.

8.7 ADTF

ADTF (Automotive Data and Time Triggered Framework) is a development and simulation environment for advanced driver-assistance systems up to autonomous driving [71]. As the name implies, the tool works with time-based data flows, based on which the behavior of parts of vehicles or whole vehicles can be controlled. Through a graphical user interface, the software enables the fusion of data from a variety of integrated sensors. Using so-called filters, these data can then be analyzed and interpreted, and in the final step, the hardware can be controlled and driving or driver-assistance functions can be performed by the vehicle.

The software is being continuously developed by Elektrobit, a global supplier of embedded and connected software for the automotive industry [75]. The German company digitalwerk holds the development and distribution license for ADTF [76]. The version discussed here is Version 3.5.0, which was released in February 2019. ADTF is available under non-free development and runtime licenses, as well as a free evaluation license for schools and universities only [1].

In the following a rough overview of the structure and functionalities of ADTF is given. Details about the individual functions and components of the framework can be found in the documentation [77].

8.7.1 Field of Application

The framework has two main application scenarios: the development of advanced driver-assistance systems [66] and the simulation of complex scenarios in Software in the loop (SIL) and Hardware in the loop (HIL) test environments [55]. ADTF is already being used by several German automobile manufacturers and TIER1 suppliers like Audi, BMW, Bosch, Daimler or Volkswagen [13, 14, 64] for the development of assistance systems, such as automatic distance control, adaptive cruise control, gesture control, adaptive bend lighting and predictive pedestrian protection. It is also highly suitable for simulating assistance systems and autonomous driving functions in virtual scenarios from the desktop [56].

In addition, ADTF can be used in combination with various other simulation tools. For example, Vladimir Entin et al. [40] used ADTF to process the virtual data, generated by the Virtual Test Drive (VTD) simulation, in the development of predictive driver assistance systems. Marc Semrau [63] examined the use of a toolchain, consisting of ADTF, VTD and Eclipse SUMO in conjunction with the ChAoS (China Assistance and other Systems) simulation framework to simulate assisted, piloted and automated driving functions, taking traffic in Shanghai as an example. In the EU research project ADAPTIVE, Michael Aeberhard et al. [32] used ADTF together with ROS (Robot Operating System), to realize autonomous driving functions. Yannick Becker, co-author of this chapter, implemented the YOLO object recognition system [31] as an ADTF filter to detect objects in a video stream in ADTF [34]. It is also used at the annual Audi Autonomous Driving Cup (AADC) [5]. In this competition, teams of students of computer science, electrical engineering, mechanical engineering or similar disciplines compete against each other in the implementation of fully autonomous driving functions. A hardware platform on a scale of 1:8, specially developed by Audi for the competition, is to be programmed, to perform various driving tasks [78].

8.7.2 Functionality

The ADTF configuration editor is the central component when it comes to implementing driving functions [79]. This is a graphical tool for creating and editing sessions. A session is a specific configuration, consisting of a Streaming

Graph and one or more Filter Graphs, that model data flows and operations on this data, to perform driving tasks [80].

The top level of a session is the Streaming Graph. It defines the data sources, the Filter Graph, that processes the data, and the data sinks, to which the processed data is sent. Data sources and sinks are graphically represented as modules with output, respective input pins, that can be connected to stream data to, respectively from, the filter graph or between each other. By default, data sources are either sensors or ADTF File Players and data sinks are actuators, which control hardware components or ADTF File Recorders. Additional data sources and data sinks can be added by the user.

The filter graph is the place, where the data from the data sources are processed and the outputs to the data sinks are calculated. A filter graph can again contain several further filter graphs, called sub graphs, in order to be able to nest and modularize functionalities. The operations are realized by so-called filters. These are encapsulated, mostly smaller programs, that each realize a certain functionality. The execution of a filter can either be triggered by incoming data, by timer triggers or by a combination of both. New filters can be programmed in C++, for example, additional algorithms for object detection. In the Configuration Editor, these are displayed as graphical elements with input and output pins, that can be integrated into the data flow by connecting them with data inports, outports or other filters. Thus, the filters allow a strong modularization of the sessions and can be easily exchanged. The filters can not only be used to manipulate the data, but also to visualize it or to generate graphical controls, such as sliders or buttons for controlling the session's behavior at runtime.

Figure 8.9 shows a sample of a streaming graph. The Universal Camera module retrieves video data from the camera, which is then streamed into the Filter Graph. Figure 8.10 shows the corresponding filter graph. Here, the camera image is read in by a streaming input and processed one after the other by two object recognition filters.

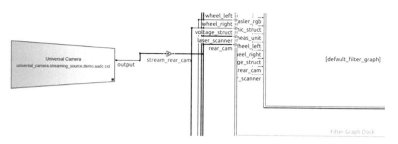

Figure 8.9: Screenshot of an ADTF-Streaming Graph.

To execute a session, the ADTF Launcher is used. It starts the ADTF runtime environment by loading and initializing the required services and plugins as well as the main processing loop with the configuration, defined in the loaded session

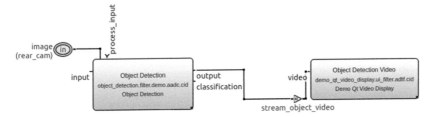

Figure 8.10: Screenshot of an ADTF Filter Graph.

file. As an alternative to the execution in the console, there is also a graphical user interface, the so-called GUI Control. This tool allows the initialization, execution and pausing of the configuration. In addition, the launcher can also be controlled via an RPC port, using a remote API.

8.7.3 Communication and Data Format

As interfaces for data exchange with external services, streaming sources serve as input and streaming sinks as outputs for ADTF. Out of the box, ADTF already offers a number of components, which can be supplemented by the implementation of custom components by the user. The already available Streaming Sources are shown in Table 8.9 and the Streaming Sinks in Table 8.10.

Table 8.9: Streaming sources included in ADTF

Component	Description
ADTFDAT File Player	Player for ADTFDAT File Format
IPC UDP Receiver	receive data from other ADTF system via UDP
IPC UDP Multicast Receiver	receive data from multiple other ADTF systems via UDP
IPC TCP Receiver	receive data from other ADTF system via TCP
IPC Host Only Receiver	receive samples via local IPC methods, using Unix Sockets or Windows Pipes
IPC SCTP Receiver	receive samples via SCTP

In addition to the various network protocol interfaces, it is also possible to exchange data via the ADTF's own ADTFDAT format. In this format, the data streams of all sensors of a vehicle can be recorded and played back time-synchronously or exported for further use. This allows to record a test drive and then use these sensor data recordings for simulation later on, for example, to reproduce erroneous behavior. Another possibility to use the recorded data is for training machine learning models like for example neural networks. This way, for example, AI for object detection in camera images or even end-to-end models for driving functionalities can be trained.

Table 8.10: Streaming sinks included in ADTF

Component	Description
ADTFDAT File Recorder	create and record ADTFDAT files
IPC UDP Sender	transmit samples via UDP to other ADTF system
IPC UDP Multicast Sender	transmit samples via UDP to multiple other ADTF systems
IPC TCP Sender	transmit samples via TCP
IPC Host Only Sender	transmit samples via local IPC Methods, using Unix Sockets or Windows Pipes
IPC SCTP Sender	transmit samples via SCTP

8.7.4 Output

Since the main purpose of ADTF is the development of advanced driver-assistance systems (ADAS), the most important output is the control of various hardware components. In addition, different visual outputs can be displayed at runtime using the Qt framework [81], to provide information about the status of the vehicle. These include, for example, the battery status, the engine speed, the video images of the mounted cameras or the measured values of other installed sensors.

During execution, ADTF offers a range of on-screen visualizations of the various sensor values, such as the video images of the cameras, point clouds of the Lidar or visualizations of the distance sensors' values, as well as the parameters of the engine or the speed of the wheels, as seen in Fig. 8.11. Here, the illustration shows, from left to right: the video image of the front camera with distortion, the video image of the rear camera and a combination of the outputs of the distance sensors in the form of filled, colored cones and of the Lidar sensor, in the form of blue points.

8.7.5 Conclusion

In summary, ADTF is an intuitive, partly-graphical programming framework for automated driver assistance as well as driving functions. Its modular design allows easy collaboration between programmers and engineers. The data and time triggers are a practical tool to control the timely execution of the functions. It also allows the integration of additional sensors as well as further software frameworks

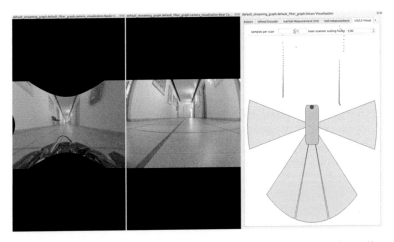

Figure 8.11: Visual Output during execution using the ADAS model car (https://www.digitalwerk.net/adas-modellauto/ – Accessed: 08-09-2020).

for special functions, such as object recognition in images or communication with external services.

The software can be used for programming driver assistance systems, up to functions for autonomous driving. Apart from this, ADTF can also be used to conveniently record data from the vehicle and various connected sensors. This data can be used, for example, when searching for malfunctions or as offline data for training AIs for autonomous driving.

8.8 Summary

In this chapter, we have presented various simulation tools for autonomous mobility and, in particular, their functionality, application scenarios and their compatibility with each other. We have also introduced ADTF, a development environment for autonomous driving functions which can also be used for simulation purposes. In some of the presented application scenarios, it became already clear, that it can be worthwhile to combine several of these tools. This allows to consider all aspects, including other intelligent agents, like vehicles or pedestrians, or specific parts of the environment, like traffic lights, road crossings or buildings, in a simulation. However, it should be noted that there are significant differences in how compatible the single programs are with other formats, when it comes to interoperability. As a platform with particularly extensive interfaces, SUMO must be mentioned here, because it is compatible with a whole range of formats, including Open-StreetMap, OpenDRIVE, Shapefiles and MATSim, which has also been presented in this chapter and also has good interoperability. A negative example is Apollo Dreamland, which only supports the Apollo API by default, and therefore can only be used with other tools, implementing it.

If the simulation tools are explicitly considered under the topic of this chapter, namely autonomous mobility, some special aspects have to be considered. As described by Allan and Farid [33], MATSim, in contrast to SUMO, can be used to simulate the specific stationary charging behavior of electric vehicles. However, one advantage of SUMO is, that, in addition to the speed and position of a vehicle, the tool also includes acceleration, which is of considerable importance for calculating consumption. Both tools, though, are not capable of mapping the effects of geographical disparities, like hills or mountains, on vehicle consumption. A further aspect, which both tools do not consider, is the underlying energy supply network, necessary for a complete simulation of electrified, autonomous traffic.

We introduced the terms microscopic, macroscopic and mesoscopic simulations at the beginning of this chapter, to be able to classify simulation tools. All tools presented in this chapter go down to the microscopic level, which means, that they consider individual vehicles during simulation. At the same time, SUMO and MATSim also allow a macroscopic view on traffic as a whole and can thus be understood as mesoscopic systems, too. Also, some simulations, such as SUMO or MATSim are more abstract, because they simulate large scenarios in a short time. Others are more visual simulations, such as LGSVL Simulator or CARLA, which use game engines, like Unity or Unreal engine, to perform near-photorealistic simulations in real time, that can be used for training machine learning algorithms for image recognition.

All presented simulation tools, with the exception of ADTF, are more or less freely available and partly open source projects. This circumstance gives the projects a larger developer base than commercial solutions and makes them openly extensible for the most part. In addition, it is important to note, that all projects, in first places Apollo and LGSVL, are under continuous development and therefore, their scope of functionality is constantly changing.

Acknowledgment

This project was funded by the Federal Ministry of Transport and Digital Infrastructure (BMVI) with the funding guideline "Automated and Connected Driving" under the funding code 16AVF2134C.

References

1. ADTF Universitaetslizenz. https://www.digitalwerk.net/adtf/universitaetslizenz/. Accessed: 24-09-2019.
2. Apollo Dreamland Website. http://apollo.auto/platform/simulation.html. Accessed: 28-09-2019.
3. Apollo open platform. http://apollo.auto/developer.html. Accessed: 20-03-2020.

4. Apollo.Auto Github Repository. https://github.com/ApolloAuto/apollo. Accessed: 02-10-2019.

5. AUDI Autonomous Driving Cup. https://www.audi-autonomous-driving-cup.com. Accessed: 24-09-2019.

6. CARLA Documentation. https://carla.readthedocs.io/en/stable/. Accessed: 24-09-2019.

7. CARLA Simulator. http://carla.org/. Accessed: 24-09-2019.

8. Dreamland Github. https://github.com/ApolloAuto/apollo/blob/master/docs/specs/Dreamland_introduction.md. Accessed: 20-03-2020.

9. Dreamland Introduction. https://azure.apollo.auto/introduction. Accessed: 28-09-2019.

10. Dreamland User Manual: FAQ. https://azure.apollo.auto/user-manual/faq. Accessed: 28-09-2019.

11. Dreamland User Manual: Quick Start. https://azure.apollo.auto/user-manual/quick-start. Accessed: 28-09-2019.

12. Dreamland User Manual: Scenario Editor. https://azure.apollo. auto/user-manual/scenario-editor. Accessed: 28-09-2019.

13. ETAS 2018. Digitalwerk startet mit neuem Vertriebspartner in das Jahr. https://www.digitalwerk.net/etas-digitalwerk-startet-mit-neuemvertriebspartner-in-das-jahr-2018/. Accessed: 24-09-2019.

14. Fahrerssistenzsysteme einfacher entwickeln und validieren. https: //www. elektronikpraxis.vogel.de/fahrerssistenzsysteme-einfacher-entwickeln-und-validieren-a-352100/. Accessed: 24-09-2019.

15. LGSVL Simulator Documentation. https://www.lgsvlsimulator. com/docs/. Accessed: 25-09-2019.

16. LGSVL Simulator GitHub Repository. https://github.com/lgsvl/simulator. Accessed: 25-09-2019.

17. LGSVL Simulator License. https://github.com/lgsvl/simulator/blob/master/LICENSE. Accessed: 25-09-2019.

18. LGSVL Simulator Python Api Documentation. https://www. lgsvlsimulator.com/docs/\#python-api. Accessed: 25-09-2019.

19. LGSVL Simulator Website. https://www.lgsvlsimulator.com/. Accessed: 25-09-2019.

20. MATSim. https://www.matsim.org/. Accessed: 19-09-2019.

21. matsim-example-project. https://github.com/matsim-org/matsim-example-project. Accessed: 20-09-2019.

22. MATSim Javadoc. https://www.matsim.org/javadoc. Accessed: 20-09-2019.

23. MATSim.org. https://www.matsim.org/open-scenario-data. Accessed: 24-09-2019.

24. Networks/Import/DlrNavteq. https://sumo.dlr.de/docs/Networks/Import/DlrNavteq. html. Accessed: 18-03-2020.

25. ns-3 — a discrete-event network simulator for internet systems. https: //www.nsnam. org/. Accessed: 24-09-2019.

26. Simulation/Output – SUMO Documentation. https://sumo.dlr.de/docs/Simulation/Output.html. Accessed: 24-09-2019.

27. TraCI/Control-related commands – SUMO Documentation. https: //sumo.dlr.de/docs/TraCI/Control-related_commands.html. Accessed: 24-09-2019.

28. TraCI/Protocol – SUMO Documentation. https://sumo.dlr.de/docs/ TraCI/Protocol. html. Accessed: 24-09-2019.

29. Via Features. https://www.simunto.com/via/. Accessed: 24-09-2019.

30. Via Licenses. https://www.simunto.com/via/licenses/. Accessed: 24-09-2019.
31. YOLO object recognition system. https://pjreddie.com/darknet/yolo/. Accessed: 30-09-2019.
32. Aeberhard, M., T. Kühbeck, B. Seidl, M. Friedl, J. Thomas and O. Scheickl. 2015. Automated Driving with ROS at BMW. ROSCon. Hamburg, Germany.
33. Allan, D.F. and A.M. Farid. 2015. A benchmark analysis of open source transportation-electrification simulation tools. pp. 1202–1208. *In*: 2015 IEEE 18th International Conference on Intelligent Transportation Systems.
34. Becker, Y. 2019. Hochautomatisiertes Fahren mit Freier Software. Master's thesis, Hochschule Trier, Umwelt-Campus Birkenfeld, 9.
35. Behrisch, M., D. Krajzewicz, P. Wagner and Y-P. Flötteröd. 2008. Comparison of methods for increasing the performance of a dua computation. Proceedings of DTA2008, Leuven (Belgium).
36. Bischoff, J. and M. Maciejewski. 2014. Agent-based simulation of electric taxicab fleets. *Transportation Research Procedia*, 4, 191–198.
37. Burghout, W. 2004. Hybrid Microscopic-Mesoscopic Traffic Simulation. pp. 14–19.
38. Chao, Q., X. Jin, H. Huang, S. Foong, L. Yu and S. Yeung. 2019. Force-based heterogeneous traffic simulation for autonomous vehicle testing. pp. 8298–8304. *In*: 2019 International Conference on Robotics and Automation (ICRA).
39. Dosovitskiy, A., G. Ros, F. Codevilla, A. Lopez and V. Koltun. 2017. CARLA: An Open Urban Driving Simulator. pp. 1–16. *In*: Proceedings of the 1st Annual Conference on Robot Learning.
40. Entin, V., T. Ganslmeier and K. Zawicki. 2009. Formale und formatunabhaengige Fahrszenarienbeschreibung fuer automatisierte Testvorgaenge im Bereich der Entwicklung von Fahrer-Assistenzsystemen. pp. 2681–2688. *In*: GI Jahrestagung.
41. Fan, H., Z. Xia, C. Liu, Y. Chen and Q. Kong. 2018. An Auto-tuning Framework for Autonomous Vehicles. arXiv preprint arXiv:1808.04913 (Preprint).
42. Fan, H., F. Zhu, C. Liu, L. Zhang, L. Zhuang, D. Li, W. Zhu, J. Hu, H. Li and Qi Kong. 2018. Baidu apollo em motion planner. arXiv preprint arXiv:1807.08048.
43. Fang, J., F. Yan, T. Zhao, F. Zhang, D. Zhou, R. Yang, Y. Ma and L. Wang. 2018. Simulating lidar point cloud for autonomous driving using real-world scenes and traffic flows. arXiv preprint arXiv:1811.07112 1 (Preprint).
44. Jérôme, H., F. Fethi and B. Christian. 2009. Mobility models for vehicular ad hoc networks: A survey and taxonomy. *Communications Surveys & Tutorials, IEEE*, 11, 19–41.
45. Helbing, D., A. Hennecke, V. Shvetsov and M. Treiber. 2002. Micro- and macro-simulation of freeway traffic. *Mathematical and Computer Modelling*, 35(5), 517–547.
46. Horni, A. and K. Nagel. 2016. Available functionality and how to use it. pp. 47–52. *In*: Horni, A., Nagel, K. and Axhausen, K.W. (eds.). The Multi-Agent Transport Simulation MATSim. London: Ubiquity Press. DOI: http://dx.doi.org/10.5334/baw.5. License: CC-BY 4.0.
47. Horni, A., K. Nagel and K.W. Axhausen. 2016. Introducing MATSim. pp. 3–8. *In*: Horni, A., Nagel, K. and Axhausen, K.W. (eds.). The Multi-Agent Transport Simulation MATSim. London: Ubiquity Press. DOI: http://dx.doi.org/10.5334/baw.1. License: CC-BY 4.0.
48. Horni, A., K. Nagel and K.W. Axhausen. 2016. The Multi-Agent Transport Simulation MATSim, pp. 3–8. 08 2016. London: Ubiquity Press. DOI: http://dx.doi.org/10.5334/baw.1. License: CC-BY 4.0.

49. Krajzewicz, D., E. Brockfeld, J. Mikat, J. Ringel, C. Feld, W. Tuchscheerer, P. Wagner and R. Wösler. 2005. Simulation of modern Traffic Lights Control Systems using the open source Traffic Simulation SUMO. pp. 299–302.

50. Krajzewicz, D., J. Erdmann, M. Behrisch and L. Bieker. 2012. Recent development and applications of SUMO – Simulation of Urban MObility. *International Journal on Advances in Systems and Measurements*, 3&4.

51. Krajzewicz, D., G. Hertkorn, C. Feld and P. Wagner. 2002. SUMO (Simulation of Urban MObility): An open-source traffic simulation. pp. 183–187.

52. Krauß, S. 1998. Microscopic Modeling of Traffic Flow: Investigation of Collision Free Vehicle Dynamics. PhD thesis, Universität zu Köln.

53. Kulathunga, G., A. Buyval and A. Klimchik. 2019. Multi-camera fusion in apollo software distribution. *IFAC-PapersOnLine*, 52(8), 49–54.

54. Lopez, P.A., M. Behrisch, L. Bieker-Walz, J. Erdmann, Y. Fltterd, R. Hilbrich, L. Lcken, J. Rummel, P. Wagner and E. Wiessner. 2018. Microscopic traffic simulation using sumo. pp. 2575–2582. *In*: 2018 21st International Conference on Intelligent Transportation Systems (ITSC).

55. Ludwig, J. 2014. Elektronischer horizont–vorausschauende systeme und deren anbindung an navigationseinheiten. pp. 223–229. *In*: Vernetztes Automobil. Springer.

56. Minnerup, P., D. Lenz, T. Kessler and A. Knoll. 2016. Debugging autonomous driving systems using serialized software components. *IFAC-PapersOnLine*, 49(15), 44–49.

57. Osinski, B., A. Jakubowski, P. Milos, P. Ziecina, C. Galias, S. Homoceanu and H. Michalewski. 2019. Simulation-based reinforcement learning for real-world autonomous driving. ArXiv, abs/1911.12905.

58. Piazzoni, A., J. Cherian, M. Slavik and J. Dauwels. 2020. Modeling sensing and perception errors towards robust decision making in autonomous vehicles. arXiv preprint arXiv:2001.11695.

59. Rieser, M., A. Horni and K. Nagel. 2016. Let's Get Started. pp. 9–22. *In*: Horni, A., Nagel, K. and Axhausen, K.W. (eds.). The Multi-Agent Transport Simulation MATSim. London: Ubiquity Press. DOI: http://dx.doi.org/10.5334/baw.45. License: CC-BY 4.0.

60. Rieser, M., A. Horni and K. Nagel. 2016. Scenarios Overview. pp. 367–367. *In*: Horni, A., Nagel, K. and Axhausen, K.W. (eds.). The Multi-Agent Transport Simulation MATSim. London: Ubiquity Press. DOI: http://dx.doi.org/10.5334/baw.52. License: CC-BY 4.0.

61. Rieser, M., A. Horni and K. Nagel. 2016. Let's Get Started. pp. 9–22. *In*: Horni, A., Nagel, K. and Axhausen, K.W. (eds.). The Multi-Agent Transport Simulation MATSim. London: Ubiquity Press. DOI: http://dx.doi.org/10.5334/baw.1. License: CC-BY 4.0.

62. Šabanovič, E., V. Z̆uraulis, O. Prentkovskis and V. Skrickij. 2020. Identification of road-surface type using deep neural networks for friction coefficient estimation. *Sensors*, 20(3), 612.

63. Semrau, M. 2018. Das ChAoS Simulations framework. pp. 31–36. *In*: Untersuchung zur Modellierung von chinesischem Fahrverhalten auf Autobahnen für den Test pilotierter Fahrfunktionen. Springer.

64. Siebenpfeiffer, W. 2014. Vernetztes Automobil. Sicherheit, Car-IT, Konzepte.

65. Sommer, C., R. German and F. Dressler. 2011. Bidirectionally coupled network and road traffic simulation for improved IVC analysis. *IEEE Transactions on Mobile Computing* (TMC), 10(1), 3–15.

66. Speth, J. 2009. Videobasierte modellgestützte Objekterkennung für Fahrerassistenzsysteme. PhD thesis. Technische Universität München,

67. Tong, K., Z. Ajanovic and G. Stettinger. 2020. Overview of Tools Supporting Planning for Automated Driving. IEEE 23rd International Conference on Intelligent Transportation Systems (ITSC). IEEE, 2020.

68. Weiss, M., S. Chamorro, R. Girgis, M. Luck, S.E. Kahou, J.P. Cohen, D. Nowrouzezahrai, D. Precup, F. Golemo and C. Pal. 2019. Navigation agents for the visually impaired: A sidewalk simulator and experiments. Conference on Robot Learning. PMLR, 2020.

69. Xu, J., Q. Luo, K. Xu, X. Xiao, S. Yu, J.-T. Hu, J. Miao and J. Wang. 2019. An automated learning-based procedure for large-scale vehicle dynamics modeling on Baidu apollo platform. pp. 5049–5056.

70. Zilske, M. 2016. How to write your own extensions and possibly contribute them to MATSim. pp. 297–304. *In*: Horni, A., Nagel, K. and Axhausen, K.W. (eds.). The Multi-Agent Transport Simulation MATSim. London: Ubiquity Press. DOI: http://dx.doi.org/10.5334/baw.45. License: CC-BY 4.0.

71. ADTF – Elektrobit. https://www.elektrobit.com/products/automated-driving/eb-assist/adtf/ – Accessed: 08-09-2020

72. Eclipse Public License – Eclipse. https://www.eclipse.org/legal/epl-2.0/ – Accessed: 27-09-2019

73. Unity 3D Overview – Unity. https://unity3d.com – Accessed: 26-09-2019

74. Overview – Unreal Engine. https://www.unrealengine.com/ – Accessed: 26-09-2019

75. Elektrobit. https://www.elektrobit.com - Accessed: 08-09-2020

76. Digitalwerk Vertriebslizenz – Digitalwerk. https://www.digitalwerk.net/digitalwerk-erhaelt-entwicklungs-und-vertriebslizenz-von-adtf/ – Accessed: 08-09-2020

77. ADTF Documentation – Digitalwerk. https://support.digitalwerk.net/adtf/v3/adtf html/index.html – Accessed: 14-04-2020

78. Autonomous Driving Cup – Audi. https://www.audi-autonomous-driving-cup.com – Accessed: 26-02-2020

79. ADTF configuration editor – Digitalwerk. https://support.digitalwerk.net/adtf/v3/guides/tools_adtf_configuration_editor.html – Accessed: 08-09-2020

80. ADTF streaming graph – Digitalwerk. https://support.digitalwerk.net/adtf/v3/guides/tools_adtf_configuration_editor.html#streaming_graph_editor – Accessed: 08-09-2020

81. Qt framework – qt. https://www.qt.io – Accessed: 24-09-2020

Deep-Learning Based Depth Completion for Autonomous Driving Applications

Christian Lanius[1,2]*, **David Hedderich**[2], **Shawan Mohammed**[1],
Gerd Ascheid[1] and **Volker Lücken**[2]

[1] RWTH Aachen University
[2] e.GO Mobile AG

9.1 Introduction

9.1.1 Autonomous Driving

Autonomous vehicles have become a hot topic in recent years due to advances in deep learning based signal processing, computational and sensor performance, and connectivity. A common way to classify the autonomy of the vehicle is the 6-level scale defined by the SAE International [15]. Traditional vehicles with no autonomous or support systems are designated a driving automation level of 0, while a fully autonomous vehicle, capable of navigating in all circumstances and not requiring human supervision, corresponds to level 5. While support systems, such as lane following, parking assistance or systems keeping a safe distance from leading vehicles are developed as standalone comfort systems, autonomous vehicles above level 3 require a holistic approach to the topic. Today, even the most advanced autonomous vehicles are not reaching level 5 autonomy.

To be able to handle such a complex system, the software is divided into disjunct tasks which can be developed and evaluated independently from each other. A common division into components, as can be found for example in Apollo [3], which is an open source autonomous driving stack, is as follows:

Maps Today, autonomous vehicles require highly accurate maps containing information about road and lane layout, as well as maps necessary for

---

*Corresponding author: christian.lanius@rwth-aachen.de

localization. These maps require significant effort to gather and are one of the main roadblocks for fully autonomous vehicles, as they introduce geo fencing as a requirement.

Localization The vehicle has to be capable to accurately localize itself in order to use to the map data efficiently, as well as perform driving maneuvers successfully. Localization is commonly done using RTK (real-time kinematic) units in development vehicles and a combination of GNSS (global navigation satellite system) and IMU (inertial measurement unit) systems, together with a sensor based localization using the map data available or a SLAM approach if no high-definition map data is available.

Perception To make the autonomous vehicle safe, it has to be capable of correctly detecting and recognizing obstacles in its surrounding. The perception module, in order to improve performance, commonly uses a multi-modal approach exploiting different performance characteristics of the sensors. Objects are commonly detected and recognized using lidars, cameras and radars.

Prediction The path of the detected objects has to be predicted for some time into the future in order to plan a safe trajectory. Prediction can be done using traditional algorithms, taking into account the previous path and the map data, or can be trained as a black box using deep learning methods.

Planning The autonomous vehicle's trajectory is planned in order to optimize the comfort of the passengers, while following the path decided by the routing approach, while making sure that no obstacles lie in the planned path.

Control Using traditional closed-loop control theory, the vehicle is kept on the trajectory planned by the planning module.

HMI Finally, the human driver should be able to interact with the vehicle, allowing engagement/disengagement of the autonomous software, as well as receive information about the operating state of the vehicle.

While this division does not allow the development of the different aspects in complete isolation, as there is always some interdependence between the different aspects, the testability is severely improved, as the surrounding aspects can be mocked in an idealized fashion. However, the final vehicle still has to be tested and evaluated in a holistic way.

9.1.2 Perception in Autonomous Vehicles

Perception is one of the key areas for autonomous vehicles. Without a rich and accurate understanding of its surroundings, the vehicle cannot safely drive through its environment. Thus, developing a reliable perception pipeline is paramount. At the same time, there are tight timing requirements: A very precise object detection is not useful, if it has a latency of multiple seconds. The objects have to be detected quickly and with a high refresh rate in order to generate accurate trajectories for them.

Today, a large variety of sensor modalities are used to detect objects. This is advantageous, as the different sensors have different performance characteristics in different situations and ranges and can thus be fused to supplement each other. State-of-the-art frameworks, such as Apollo [3], use multiple lidars and cameras, as well as commercial radar solutions for object detection. At night for example, cameras provide little information, while lidar and radar transmit their own impulses and so are not impacted. Lidar might perform badly in situations where the transmissivity of the air is low, such as rain or fog. In general, the resolution of lidars and especially cameras is higher than commercial radar solutions, enabling a more accurate localization of objects in uneven road conditions, such as at the foot of a hill.

Figure 9.1 shows the perception pipeline for Apollo: It can be split into two disjunct problems: Detection and recognition of traffic lights as well as detection, recognition and tracking of objects. Traffic light detection is done solely using cameras. To reduce the computational cost, high quality annotations for the location of traffic lights in the map material is used to obtain an expected location of the traffic light. The traffic light detection thus only has to be executed if the vehicle is moving into the vicinity of an intersection. On the region of interest, obtained by projecting the map information into the camera frame using the ego vehicle's location, a region of interest is obtained. On this region, a YOLO (you only look once) based object detector [33] is applied to obtain the actual location of the traffic light. Afterwards, a CNN based classifier is used to obtain the current state: Red, green, yellow, black. The black state is given if the classifier is not sure. This might be the case if the traffic light is turned off. The vehicle then applies a simple temporal filtering: After a red light, red/yellow has to follow,

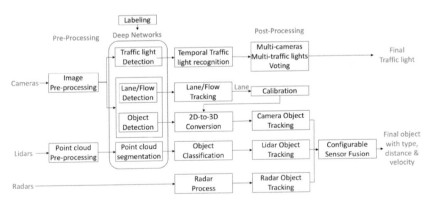

Figure 9.1: The complete perception pipeline for Apollo, a state-of-the-art AD framework. Reproduced from [2].

but not green etc. Only if the detected state is constant over multiple frames, the published state changes. If multiple cameras are available, their results are voted on by the algorithm, and the majority state is used.

Object detection is a more challenging problem: The location, type and identity of multiple objects has to be correctly estimated, with a much larger solution space compared to a traffic light. There exists a wide variety of fusion algorithms, such as [25], dealing with varying update rates of the different sensor modalities and their accuracy. Tracking is commonly done in a two-stage tracking-by-detection style, where new detections for each sensor modality are matched to existing tracklets according to a cost function which is optimized using algorithms such as the Hungarian method [21]. Tracking is then refined with a new assignment between existing tracks and the fused object from the fusing algorithm. The cost function commonly uses features such as similarity in appearance, distance, velocity etc. In the following, we will present, as an example, the implementation by an open source, state-of-the-art AD (autonomous driving) framework.

The camera object detection in Apollo uses a deep learning model to detect 2D bounding boxes for the obstacles, which are then converted into 3D. In order to obtain this mapping, lane boundaries, horizon and vanishing point are estimated using a deep learning model. Using the camera calibration, the distance of the objects can then be estimated.

The lidar object detection is the most sophisticated. It first removes points outside a region of interest, which is obtained by removing all points which are outside the road area. The remaining points are then projected into a regular grid and eight measures are computed for each cell: Highest point in cell, it's intensity, mean height and intensity, number of points, angle and distance of the cell to the vehicle location and finally a binary value if the cell is occupied. Using a deep learning model, for each cell, objectness, positiveness, object height and class and offset to the center of the object are estimated. Using the center offset prediction, a directed graph is created, which can be divided into clusters using the Union Find algorithm [11]. Finally, small objects, and those where the height does not correspond to the detected objects height, are rejected. As the clustered cells are only convex, a minimum bounding box is drawn around it. The object class is given by averaging over the prediction from the deep learning model.

9.2 Fundamentals

9.2.1 Deep Learning

Until a few years ago, most algorithms for problems from all kinds of domains required expert domain knowledge and a strong background in statistics, signal processing and system design to create robust classifiers, detectors and segmentors. For computer vision task for example, a pipeline based on filtering, edge detection and clustering was very common.

The final classification was often done using meticulously designed features, encoding the priors the system architect is aware of in the problem setting, which are then grouped using supervised machine learning techniques such as support vector machines, random forests or *XGBoost*. Hand-engineering the features is necessary, as the input space is high-dimensional but very sparse; and thus has to be reduced to a lower dimensionality to be tractable.

This hand-engineered feature extraction has been replaced with deep learning in recent years. Major advances in computational capabilities, tooling and architectures have led to deep learning, where the feature extraction itself can be learned. Today, GPUs (graphics processing units) and ASICs (application-specific integrated circuits), together with complex and optimized libraries, have accelerated design, training and development massively. Recent research developments, such as residual connections, momentum based optimizers and learnable normalization have led to bigger and deeper models which still can be trained consistently. This leads to even non-domain experts to be able to build models providing a strong baseline for a large variety of tasks.

However, deep learning introduces a new set of problems and challenges. Two main barriers of entry are building up and can become a hurdle difficult to overcome for new players in the space: Data and compute. In recent years, the performance of models can be always be improved by providing it with more labeled data. As data has to be labeled by hand, creating a new, large and diverse dataset has become a very challenging task. Therefore, a large and active field of research is using techniques to reduce the need for large amounts of labeled data, such as knowledge distillation, teacher-student learning or transfer learning. However, even ignoring the access to manpower for labeling and gathering datasets in the required size requires enormous resources. Thus, companies like Google and Facebook are at the forefront of deep learning with their massive user bases.

While consumer-grade hardware today, especially GPUs for general purpose computing using NVIDIA CUDA [31], are many times more powerful than just 10 years ago, the trend to bigger and deeper models lead to big players developing their own hardware to better support their deep learning efforts, such as Google's TPUs [18] and Tesla's Autopilot chip, TRIP [22]. This can lead to state-of-the-art models, while having the training set and model specifics available, to not be reproducible. While this case is common in other disciplines of science (nobody is able to reproduce the experiments carried out at CERN), this is a departure from deep learning in recent time, where the compute playing field was more equal.

An additional problem introduced by deep learning is explainability: A traditional algorithm is straightforward to debug or perform error analysis on in case of a major issue. Deep learning on the other hand is still in it's infancy in this regard. This aspect becomes important in safety-critical applications, such as autonomous driving. Being able to analyze a situation after it happened is important in cases where the number of occurrences of similar situations is low, so the few datapoints available have to be exploited to the maximum, which cannot be guaranteed with deep learning algorithms.

9.2.2 Sensors

Camera

Today, cameras can provide a high resolution image of the surrounding area. Using different optics, the field of view can be setup in a large design range and the frame rate and exposure can be adapted according to the environment.

From a technical point of view, a camera produces an ordered 2D representation, with three color channels. Optics and other distortions can be accurately measured and removed from the image. Cameras used in autonomous vehicles commonly feature resolution of 1 to 2 megapixels and relatively low framerates of 15 fps. These values are mainly chosen in order to limit the compute necessary, as cameras can be commonly built with 10 times the resolution and 4 times the framerate.

In autonomous vehicles, cameras are mainly used for 3 different reasons: They are the only way to properly discern text, colors and printed shapes, enabling the vehicle to recognize traffic lights and signs. As computer vision is a topic well studied, and has seen major breakthroughs in recent years due to deep learning, object detection and recognition using cameras can be done with off the shelf algorithms. Finally, VSLAM (visual simultaneous localization and mapping) is an active field of research, aiming at using camera and images to localize the vehicle.

Lidar

Lidar sensors provide a direct and straightforward way to obtain non-dense point clouds. While there a different approaches to building lidar sensors, they all share that they are based on lasers outside of the visible spectrum. The most common approach, popularized by Velodyne, uses a spinning sensor with a number of detectors aligned on a vertical axis. A pulse of light is sent out, reflected at objects and the runtime until measurement is used to determine the distance from the sensor. By stacking the detectors, the vertical perceptive field is determined. The main figure of merit for a lidar is the vertical perceptive field, the horizontal and vertical angular resolution and the update rate. For high-end lidars, like the Velodyne Alpha Puck, the vertical angular resolution is increased close to the ground plane, so that the horizontal resolution is not constant [1]. It features 128 layers in total, with an angular resolution of 0.1°, and a total vertical FOV of 40°. Due to the spinning nature, it has a horizontal 360° view with an angular resolution of 0.1°. These high performance characteristics are coming at a cost though.

Spinning lidars, while being state-of-the-art, have some major downsides: They are quite expensive, though the price levels are changing slowly. To be able to provide a full 360° view, they have to be placed in an extruded position. Otherwise, multiple sensors have to be combined to provide such a large FOV.

As they are spinning with 600 rpm and more, while providing high angular resolution, there are stringent requirements on motor control, accuracy and a reduction of car movement impacts. Thus, recent development in lidars focuses on using either MEMS based mirrors or developing true solid state lidars, similar the approach used in modern CMOS camera sensors. While being less sensitive to mechanical influences and being able to be placed less prominently, for example behind the wind shield, by design they feature smaller horizontal FOV compared to mechanically spinning lidars.

9.3 Related Work

9.3.1 Dataset

The training and evaluation of the proposed method was carried out on the so-called KITTI dataset. It is a large scale urban and semi-urban dataset, containing a large variety of modalities, gathered from a moving vehicle [12]. The data available are stereo gray and RGB images, 64-layer lidar data and a GPS/RTK unit for accurate localization, as well as exact calibration. The data was gathered in 2012 in Karlsruhe, Germany, by researchers of the KIT (Karlsruhe Institute of Technology). The raw data is divided into different drives of varying length, most of which are short snippets, normally 70–200 consecutive datapoints gathered by the sensors, but some are multiple thousand frames long. Over the years, more and more ground truth data has been gathered [12, 13, 30, 38] to provide a variety of challenges, including public leaderboards. The work presented in this chapter attempts to solve the *Depth Completion* challenge.

The benchmark task consists of dense depth map prediction given a sparse depth image, which is generated using a lidar. As an auxiliary input, synchronized RGB images are available as well. One of the downsides is that the ground truth is not available for all points, such as the sky or outside the vertical FOV of the lidar, where no annotations are available. This results in most models severely degrading in performance for the upper 30% of the image. If there was a conflict of the different methods used to gather the ground truth data, the area does not contain information either. The ground truth is generated by aggregating a number of consecutive lidar frames, fusing them by applying semi-global block matching, taking the vehicle movement into consideration, and finally manually verifying and correcting wrong patches. Finally, the point clouds are projected into the image frame. Due to this semi-automatic nature, a large number of frames are available, around 94,000. On the same dataset, simply by ignoring the sparse depth completely, depth estimation only from images/videos can be explored as well.

9.3.2 Methodology

In this section, related work is presented, upon which the proposed method in Section 9.4 is built upon.

UNet

Image-to-image translation tasks have early been identified as a possible application field for convolutional networks as they are can be shift invariant, which is a desired property for image translation tasks. At the same time however, they have to capture high level context from the entire image. Previous work used a sliding window approach, where the perceptive input size of the model is smaller than the input image [8], or where classification is preceded by a region extraction scheme [14]. These models miss out on the contextual information available in the image which is not inside the perceptive field. These approaches could be built to take the entire image into consideration, but the computational and memory limits restrict the resolution of the input image, and thus the fidelity of the output. Training such a fully convolutional model also requires a larger dataset, compared to models using pooling operations to increase their receptive field.

First proposed for biomedical image segmentation, UNets attempt to solve image segmentation more efficiently, being able to take into context spatially distributed, highly abstract context and local information to generate accurate segmentation for cell segmentation [34]. The proposed UNet architecture is shown in Fig. 9.2. The name-giving U shape is directly apparent. It uses blocks of convolutions, followed by downsampling to generate a low resolution, high context vector (shown in the bottom center). This state is then upsampled stepwise into a high resolution segmentation map using deconvolution layers, shown in green. The term "deconvolution layers" might be an unfortunate choice, introduced by the authors in [35], as the layer does not perform deconvolution

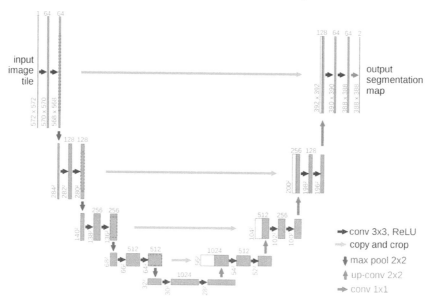

Figure 9.2: The proposed UNet architecture. Reproduced from [34].

in its traditional sense. If convolution in the time domain is $g(t) = f(t) * h(t)$, then in the frequency domain it becomes a multiplication, thus a deconvolution as an inverse of convolution would be equivalent to finding such an $f(t)$, that $F(f) = G(f)/H(f)$. Instead, deconvolution, sometimes also called transposed or up convolution, is a convolution with a stride less than 1. It essentially is a learnable upsampling step, where the weights of the upsampling are not fixed, but learned through backpropagation.

This U shape of networks is also common for autoencoder networks, which learn an identity mapping, but by forcing the image through the low resolution stage extract an information dense representation. This is motivated by information theory and is also called the information bottleneck and can be used to learn a compressed encoding of the input, similar to principal component analysis (PCA). However, this is in general not a lossless encoding. To preserve well defined borders between objects, skip connections are introduced by the UNet architecture. Each upsampling stage has a corresponding downsampling stage, whose output dimensions are the same as the input dimension of the upsampling stage. By concatenating the upsampled, low resolution, high information, representation with the high resolution, lower information encoding from the downsampling stage, the upsampling stage can generate a high resolution output, which can take both aspects into consideration.

Today, this architecture is used in a variety of models for all kinds of image-to-image tasks, such as image segmentation [43], domain translation [17] (pix2pix uses a UNet architecture for the generator), or human pose estimation [36], which uses multiple stages of hourglasses, which are very closely related to UNets, to refine keypoint estimates.

In summary, UNets provide a powerful and efficient architecture to train image-to-image translation tasks. The trained models are able to access high resolution information, as well as low resolution, highly abstract representations of the input. They provide state-of-art performance for this class of problems.

Depth Completion

KITTI provides a semi-dense ground truth to evaluate depth estimation tasks on. The dataset is big enough to be able to learn on it using traditional supervised methods (which might have to be modified to handle the sparseness if necessary). State-of-the-art performance is obtained in this way [37]. However, this abundance of high quality labels is not always available. Thus, recent work went into methods for allowing models to learn with only little or no ground truth data.

In 2017, the "Structure from Motion (SfM) Learner" [44] was introduced at CVPR (IEEE Conference on Computer Vision and Pattern Recognition). Structure from motion is a very old topic and has been published about in the late 70s. The underlying idea is to recover 3D information from a camera, creating 2D images, moving through a scene. For this, the camera's position and orientation (PnP, pose and position) have to be obtained. This is normally done by tracking feature points, obtained using algorithms such as SIFT [24] or SURF [4], through the sequence to then reconstruct the camera motion. After solving the PnP problem, 3D aspects such as disparity can be calculated, assuming that the found transformation is accurate. This way, the 3D structure can be obtained from motion, giving the method its name.

As deep learning attempts to remove the manual feature engineering and tuning necessary for such an approach, SfM Learner, as the name suggests, attempts to learn the motion of the camera and the depth simultaneously, given an image sequence. For every frame I_t, the model utilizes the next and the previous frame I_{t-1}, I_{t+1}. For every point in I_t, the depth estimation network estimates the depth using only that frame. Using all three frames, the pose estimation model predicts the transforms $T_{t \to t-1}$ and $T_{t \to t+1}$, every 3D point in I_t can then be projected into I_{t-1} and I_{t+1}. As the loss term has to be differentiable w.r.t. the inputs, bilinear interpolation is used to obtain the true image value at the projected position. This approach requires that the scene is static and there is no occlusion between the different frames. To accommodate these limitations, the model additionally learns a confidence mapping E, which allows it to weight the pixels differently in the loss function. Thus the loss the network attempts to minimize is:

$$\mathcal{L}_{photo} = \sum_s \sum_p E_s(p) \left| I_t(p) - \hat{I}_s(p) \right| \tag{9.1}$$

where p are the pixels in the source image s and t are the indexes of the target images, i.e. the previous and next image. $\hat{I}_s(p)$ is the intensity of p where I_s is warped into I_t, according to the estimated transform. s is a downscaling factor. As the bilinear interpolation only takes into consideration the surrounding pixels, the warping has to be accurate down to 2 pixels. To alleviate this, the warping is evaluated at multiple scales.

This function has a simple solution for which the loss will always become 0. If the confidence of the model is 0 for all points, the loss trivially becomes 0. To counteract this, a regularization term $\mathcal{L}_{reg}$ is introduced. The authors introduce this term as the binary crossentropy to a constant 0 and argue that this makes the model try to minimize the photometric loss, while giving it some slack to account for the model limitations.

The final loss term is a smoothness constraint $\mathcal{L}_{smooth}$. As at most locations, the gradient of the depth is expected to be constant, i.e. the scene is roughly made up out of flat planes in 3D space. Thus, this term is chosen to be the L1 norm of the second derivative of the estimated depth image. The final loss term which is minimized is a weighted sum of these three aspects.

The presented paper [26], which is based on previous work from the same authors [27], attempts to create dense depth estimates using a self-supervision framework very similar to the one proposed in the SfMLearner style networks. However, there are some key differences we will highlight. Note that this paper has been the base upon which the proposed method in Section 9.4 is built.

For the proposed self-supervised framework, as shown in Fig. 9.3, the authors use a three-part loss function as well. However, they drop the confidence estimation network and the related regularization loss. As they have a sparse depth map available as an input, they enforce that model learns an identity mapping for those points. The sparse map is encoded as having depth values where available and is 0 otherwise. No sparse convolutions were used and the model is expected to learn this implicitly. If d_1 is the sparse depth (projected into the RGB image 1, for which we want to estimate a dense depth) and P is the set of all points in the image, then the depth loss is:

$$\mathcal{L}_{depth} = \sum_p \left(\left\| d_1(p) - \hat{d}_1(p) \right\|_2 \right) \forall \; \{p \in P | d_1(p) > 0\} \tag{9.2}$$

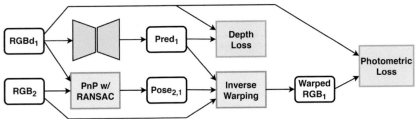

Figure 9.3: The proposed training architecture for the self-supervised depth completion. The smoothness loss is not shown in this figure. Reproduced from [26].

Outside of this additional loss, the loss definition is the same between the two approaches. Due to the inclusion of the sparse depth as input, the model performs much better than the SfMLearner type of models, even considering the improvements made to it in [6, 29]. The estimation of the pose and position using traditional structure from motion methods, has replaced the movement estimation model.

While deep learning as an approach has been applied to a large variety of problems in different disciplines, often outperforming existing state-of-the-art, a well-crafted traditional algorithm with pre-processing can sometimes compete with deep learning models, with a fraction of the compute necessary. Originally implemented for a 3D object detection algorithm [20], a simple image processing algorithm, the authors called IP-Basic, only taking the sparse depth into consideration, has shown very promising results for depth completion [19].

The algorithm consists of seven straight forward image processing steps: The point cloud is inverted, i.e. all valid values are subtracted from 100 m, as the library [5] used to implement the algorithms takes larger values over smaller values if necessary. As invalid values are 0 m, closer objects take precedence over farther values, which take precedence over pixels with no input value. The sparse depth is then dilated using a diamond shaped kernel of size 5 × 5. The size and shape of this kernel was a hyperparameter and was chosen so that it optimizes the MSE of the resulting depth estimate. Small and medium holes are closed in a third step using a binary closing operation and dilation of the missing values mask. As the sparse depth is only available for the lower two-thirds of the image, the values are extended upwards to obtain an estimate for the entire image. To fill the remaining holes, a large scale dilation of size 31 is applied. To remove unwanted outliers, a median blur is applied to the image, followed by a gaussian kernel, both using a kernel size of 5. Finally, to recover the original depth, the image is inverted again.

A comparison for the presented methods is shown in Table 9.1. Additionally, the state-of-the-art algorithm [37], which uses the ground truth for learning, is shown as well. Because of the inclusion of ground truth during training, it outperforms all self- and unsupervised methods.

Table 9.1: Comparison of the presented methods. The root mean square error, as main metric of the KITTI depth completion challenge, is reported

Method	Input	RMSE [m]
[6]	RGB	4.7503
[19]	D	1.2885
[26]	RGBD	1.3849
[37]	RGBD	0.7362

Semantic Segmentation

Semantic segmentation is the task of generating a label for every pixel in an image. In contrast to instance segmentation, the goal is not to generate sets of pixels corresponding to distinct objects. Common classes for this task, e.g. the classes in the KITTI semantic segmentation challenge, as well as the Cityscapes dataset [9], are roads, pedestrians, different types of vehicles, vegetation or the sky.

The state-of-the-art in this field is dominated by deep learning methods, as the problem can be formulated as a pixel-wise classification problem, which is a field well understood. Obtaining a large-scale dataset is a challenging task and, due to the large manual effort, most datasets only consist of a "small" number of labeled frames. The KITTI semantic segmentation task e.g. only consists of 400 labeled frames, split into training and validation set. Thus, a lot of effort is spent on reducing the amount of manually labeled data required.

For the approach presented in Section 9.4, a semantic segmentation is needed. To obtain this segmentation, the state-of-the-art model, as presented in [43] and implemented by the original authors, is used. It's main contribution is the extended use of image and label propagation to increase the dataset size, as well as a relaxation of the labels in boundary regions.

Label propagation exploits the temporal consistency: By matching features between the labeled and consecutive frame, and then applying the found transformation to the label, an approximate label for the consecutive frame can be found. To alleviate the problem of artifacts due to misalignment, instead of only propagating the label, the original image is propagated as well. If no misalignment existed, this step would not be necessary. The observed propagated image is aligned better to the propagated label than the consecutive frame.

After the joint propagation of the images and labels, due to "due to ambiguous annotations or propagation distortions", boundary regions become less clearly defined, i.e. their entropy has increased. In these boundary regions, i.e. where a pixel has a different label than any of its neighbors, the probability that the pixel is in any of the classes in its neighborhood is maximized. If all surrounding pixels share the same class, this trivially reduces to maximizing the probability that the pixel has the one hot encoded value in the ground truth. Otherwise, if N are all classes in the pixels neighborhood, the loss to minimize is defined as

$$\mathcal{L} = -\sum_{C \in N} P(C) \tag{9.3}$$

The authors use the proposed method to generate a larger dataset with relaxed boundary conditions and use it to train a DeepLabV3Plus [7] architecture with a WideResNet38 backbone [40]. The model outperforms previous models and is, at time of writing, the best performing model on the semantic segmentation task of the KITTI dataset with an IOU score of 72.82.

Group Normalization

Until the introduction of normalization layers, deep learning models where prone to sensitivity w.r.t. the initialization, requiring small learning rates and long learning cycles. To improve generalization, dropout was used to lessen the impact of single neurons on the output. It has long been known, that a normalization of the input increases convergence speed. Inside of multi-layer networks, this normalization was impossible, as the outputs were not available at training time and change over the training. In contrast, the mean and variance of the dataset can easily be calculated and thus can be normalized. A large range for the output values of internal layers lead to a large number of neurons not having any impact on the output. For all non-linearities with vanishing gradients in certain regions, backpropagation results in very small updates: For example ReLU layers have no gradient for negative input values. Thus, large negative inputs to the non-linearity result in no weight update, effectively decreasing the learning speed of the network and leading to "dead" subnetworks which have no impact on the output.

This changed when batch normalization was introduced in 2015 [16]. Using the minibatch statistics, it learns parameters using normal backpropagation to output a vector where every channel is normalized over the minibatch. For a minibatch of size m, the mean and variance of each channel are calculated. Then, each channel is normalized using these values. Finally a linear, learnable transformation is applied. Otherwise, symmetric, non-linearities would be constrained to linear outputs. During inference, any influence of other samples in the batch is undesirable: The output of the network for each sample should only depend on the respective input. Thus, during training a running average over the training datasets expected value and variance for each layer is being kept track off. Thus after training, we obtain an estimate for both values, which is then frozen and used for the scaling during inference.

Batch normalization improved deep learning massively, it allowed faster training, more aggressive learning rates and decreased sensitivity to initialization. However the greatest issue of batch normalization is the requirement of relatively large batchsizes: If the number of samples in a mini batch is small, the estimated expected value and variance might differ significantly compared to the true values over the entire dataset. A large batchsize thus became a requirement to properly train deep networks. It increased the memory requirements to train models and lead to a major impact of batchsize as a hyperparameter. The larger the batchsize, the better the estimated values and the better the generalization and performance of the model.

With increasing complexity and depth of models, this led to research into other types of normalization. The most popular are layer normalization, instance normalization and especially group normalization [39], which is a combination of the previous two. Figure 9.4 shows a visual comparison between the methods: The approach for all methods is the same: Normalize the output over the dark blue squares, batch normalization normalizes for every channel, instance norm for every sample, and the entire layer's output is normalized per sample with a layer norm layer. It can be seen how group norms are a mix between the previous two: The output of a layer is divided into groups, which are then normalized independently from each other. This approach is motivated from the group wise aggregation and normalization in more traditional feature extractors like SIFT and SURF.

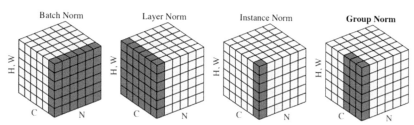

Figure 9.4: A visual comparison of different normalization approaches. N is the batch dimension, C is the channel axis and (H,W) is the spatial axis. Reproduced from [39].

While there is no real theoretical proof, the authors show that group normalization removes the sensitivity from the batchsize inherent in batch normalization. Thus, group normalization promises good results while using small batchsizes, reducing the memory requirements and allowing deeper and more complex models on the same hardware, with no accuracy trade-off. While a large batchsize still stabilizes stochastic gradient descent methods, its impact on final accuracy is thus far smaller compared to models using batch normalization.

Due to limitations in computing and memory resources, this novel normalization layer is used in the method proposed in Section 9.4.

9.4 Proposed Method

As presented in Section 9.3.1, there exists a large amount of data available for a variety of computer vision tasks. In the following, we will consider the challenge of creating a dense depth map from a single RGB image, synchronized with sparse lidar data. Especially in the autonomous field, this combination of sensors is common and the resulting dense depth maps can serve as foundation for a variety of applications.

9.4.1 Motivation

For a large scale deployment of autonomous vehicles, the cost is going to be a major factor. Few customers are going to be willing to spend six-digit numbers on the hardware required to make an autonomous vehicle safe. Even for companies such as uber and lift, expensive hardware stacks are an obstacle. They drive up the cost per vehicle, especially in the case of accidents, decreasing their margins and competitive advantage compared to human drivers.

As the lidar is one of the main contributors of cost, replacing it seems to be the most straightforward way to reduce costs. Some companies, such as Tesla, do not employ one at all. At the same time, the major advantages of lidars lead to their ubiquitous use in large parts of the industry: They generate high accuracy, 360° depth information, in real world coordinates. Especially this last property is very desirable: Even up-to-scale depth information from e.g. cameras using disparity maps, does not allow the vehicle to accurately gauge the absolute distance to possible obstacles.

At the same time, the field of computer vision is developing extremely quickly since the deep learning revolution in the first half of this decade. A lot of approaches for a large variety of problems are investigated in this aspect. Therefore, autonomous vehicles can employ this research to improve their performance.

The main difference between lidar and a camera image is the ordering: A camera image is an even grid of samples, where for each point a value (as a triplet of red, green and blue) is available. In contrast, a lidar's output is an unordered point cloud: It is simply a list of points and their associated reflectivity. While these points can be mapped into the reference frame of the image using the extrinsic and intrinsic parameters of the setup, there is no guarantee that every pixel in the image has an associated point in the point cloud. Indeed, modern cameras with a FullHD resolution (1920 by 1080 pixels), and a common field of view of 69° by 42°, this results in an angular resolution of $0.035°/px$. Spinning top lidars on the other hand generally have different spacing in the horizontal and vertical direction, which is still not as high resolution as the camera's. The flagship lidar from Velodyne for example has a resolution of around 0.1°, which is still worse than common cameras by a factor of 3, completely ignoring modern cameras boasting up to 15 million pixels, as these resolution increase the computational burden and bandwidth requirements.

As the lidar's main differentiating feature is the number of layers, i.e. the vertical field of view and resolution, to reduce cost the easiest way is to downsize the lidar. While a 16-layer automotive high-end lidar costs around 4,000 USD, the variant with four times the layers is not four times as expensive but rather 19 times, coming in at around 75,000 USD. Therefore, a combination of cameras with lidars is a straight forward solution to fuse the sensors in order to provide depth estimates in the high resolution of cameras, coupled with the real world scale and accuracy of lidars.

9.4.2 Proposed Architecture

We propose a self-supervised, semantics-aware deep learning approach to generate dense depth maps from single RGB images and sparse point clouds.

We use the data made available in the KITTI depth completion task. We use the provided data split for dividing the data into a training and a validation set and all values we report are obtained from the validation set, as the ground truth data for the test data is not available.

As a first preprocessing step, we generate semantic segmentation masks for all training images. While we initially considered to provide these masks as input, we dropped this idea, as the generation is too slow for real-time performance. Thus, they are only used for the training step and are not needed during real-life application of the model.

Model ensembling has long been known to improve the performance of classifiers and regressors [10]. Using the confidence estimates we obtain from our deep learning model, we filter the sparse point clouds to remove outliers before using a depth estimation model, implemented using traditional computer vision techniques [19], to obtain a second depth estimate. We then fuse the two depth maps to outperform, in terms of RMSE, both independent models. A graph shown the processing pipeline is shown in Fig. 9.5.

9.4.3 Metrics

We report the metrics as proposed by [38] and [26], as is commonly done for this problem:

RMSE The root mean square error, in mm.

MAE The mean average error, in mm.

SiLog The scale invariant log error, as proposed by [38].

Squared Rel The squared relative error.

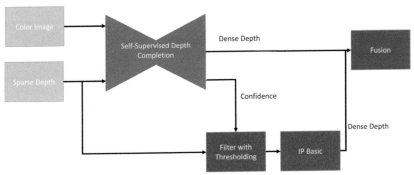

Figure 9.5: The proposed model to handle depth and confidence estimation in an unsupervised fashion.

IRMSE The inverse root mean square error, essentially the RMSE for the disparity map, up to a constant factor, in km^{-1}.

IMAE The inverse mean average error, essentially the MAE for the disparity map, up to a constant factor, in km^{-1}.

Abs Rel The absolute relative error.

δ_1 The percentage of pixels, where the ratio between prediction and ground truth is between 0.8 and 1.25.

δ_2 The percentage of pixels, where the ratio between prediction and ground truth is between 0.82 and 1.252.

δ_3 The percentage of pixels, where the ratio between prediction and ground truth is between 0.83 and 1.253.

9.4.4   Deep Learning Model

We implement a vanilla UNet architecture for our model, using early fusion. The detailed model is shown in Fig. 9.6. The in- and outputs are shown in yellow. Dashed connections represent skip connections and circles correspond to concatenation along the feature axis. A single stage is used for feature extraction of each input modality before concatenating the resulting feature maps. It consists out of a 3 by 3 convolutional filter, followed by normalization and non linearity, shown as green blocks in the figure. The resulting tensor is downsampled in four stages, successively reduced in spatial resolution, while increasing the number of features. Each of the stages, shown in orange, is implemented as a stack of two typical residual blocks. Finally, using only one filter, it is downsampled to a low resolution, high dimensionality representation. The image is in total downsampled by a factor of 32, and increases its number of channels from four (RGB + D) to 512.

Using transposed convolution blocks, in blue, the image is the successively increased in dimension again: Four upsampling blocks increase the size and decrease the dimensionality again. Each block consists of a strided convolution,

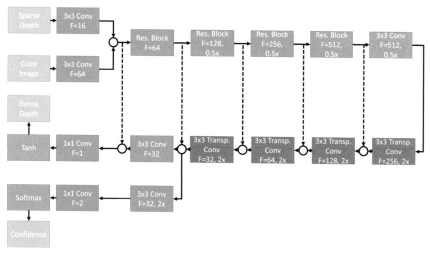

Figure 9.6: Proposed model architecture.

followed by normalization and a non-linearity. As is done with UNet architectures, each upsampling block's output is concatenated with the matching down-sampling block before being fed into the next stage, in order to combine low resolution, high context information with high resolution, low context data.

The result is then fed into two separate heads of the network, which predict confidence and depth for every pixel. We apply a tanh nonlinearity on the last layer to restrict the output to the range of 0 to 100 m. The confidence output is using two channels, calculating the pixel-wise logarithmic softmax. These non linearities are shown in grey. We are mainly inspired from the work presented in Section [26], however, we made some different choices in the design of the model.

As alluded to in Section 9.3.2, the large downside of batchnormalization is the requirement for large batch sizes and sensitivity to batch size in general. Thus, we replace the batchnormalization at all instances with group normalization. As our hardware, while being high end consumer grade, is not big enough in fit large batch sizes, we hope to keep performance at a comparable level to the baseline while reducing the batch size. In the reference, the authors used a batch size of 8, while we only use one of 2.

The verdict is still out there on the use of more complex activation functions compared to simple ReLU functions. Most papers seem to suggest very little difference between ReLU and leaky ReLU activations in deep networks [28, 41], so we use the more traditional ReLU activation functions, due to the decrease in computational need and number of hyperparameters to choose.

Another choice we had to make to reduce the computational requirements was to half the number of channels in the upsampling part of the network, as well as reduce the number of convolutional layers in the downsampling stage: The original authors used blocks as implemented in the ResNet32 architecture, while we use blocks as used in the ResNet18 case.

Finally, the included non linearity on the depth output is added by us. The original paper did directly use the linear output. However, this necessitates the model to learn the valid value range itself. By applying a non linearity at the end, which compresses all values to the valid data range, we can move all values to the valid range. As some of the metrics we use are only defined for values greater than 0, this also gets rid of invalid results which will propagate. The downside is the decrease in sensitivity for close and far regions: For the network to output a very small value, the last layer has to output a very large negative value, for values close to the maximum distance, the value fed into the tanh layer has to become very large.

9.4.5 Loss Function

Depth Loss

As input for our model, we consider images, as well as a sparse depth, while we want to predict a dense depth map. In an idealized fashion, all points, for which we have input samples available, should have the exact same value. Thus for those points, we want the model to learn an identity mapping.

Given that the metric we care most about, and which we use to rank our models is the root mean square error, choosing the L^2 norm over the image, at points where we have samples available is a straight-forward decision. Thus, in a first formulation, our depth loss can simply be expressed as

$$\mathcal{L}_{depth} = \sum_p \left(\left\| d_1(p) - \hat{d}(p) \right\|_2 \right) \forall \{p \in P | d_1(p) > 0\} \tag{9.4}$$

which is the same definition as in Equation (9.2), as proposed by [26]. However, not all lidar points are necessarily valid: They might be invalid due to synchronization issues or fast moving objects. Indeed, the problem occurs in the KITTI dataset, being most pronounced at the borders of objects at the side of the screen, in fast moving environments, such as highways, and for very narrow objects like poles. Two examples of these cases are shown in Fig. 9.7. The loss function in Equation (9.5) is not well equipped to handle this error.

As alluded to before, we define a confidence measure to the output of our model. We hope that the model learns to ignore lidar points, if they are most likely false, i.e. it has a low confidence that the value is used. If a point is false, it's error in the identity mapping should be ignored. We can achieve this by weighting every pixel, which has a lidar point, by this confidence. This turns Equation (9.5) into

$$\mathcal{L}_{depth} = \frac{\sum_p \left(\hat{c}(P) \left\| d_1(p) - \hat{d}(p) \right\|_2 \right)}{\sum_p (\hat{c}(p))} \forall \{p \in P | d_1(p) > 0\} \tag{9.5}$$

Figure 9.7: Two examples for KITTI exhibiting the same problem of the raw lidar data containing invalid samples. The vehicle on the left hand side has points on its hull (true value should be blue), which are yellow, i.e. further away. The right side of the cyclist is also estimated to be further away than the true distance, shown by the greenish points.

i.e. the weighted sum of the mean square error. This is inspired by a similar formulation used in [6], where the same approach (without dividing by the sum over $\hat{c}$) was used to account for inaccuracies in their photometric loss calculation.

This formulation allows us to theoretically handle wrong lidar measurements. However, like this, the function has no impact: By simply learning a constant value for the entire image, it only scales the loss value. Thus, we need to penalize the function to choose values other than 1: We do this by introducing a regularization loss, again as proposed by [6].

We use the mean square error to 1 as our regularization loss, as it showed good performance, and the observed probability distributions matched the bimodality we wanted to achieve: A high peak at 1, where the points are true and a second, lower peak at 0, which were the samples we want to discard. The final loss function for our lidar points can thus be expressed as:

$$\mathcal{L}_{\text{depth}} = \frac{\sum_{p}\left(\hat{c}(p)\left\|d_l(p) - \hat{d}(p)\right\|_2\right)}{\sum_{p}(\hat{c}(p))} + \sum_{p}\left(\left\|1 - \hat{c}(p)\right\|_2\right)$$

$$\forall \, \{p \in P \,|\, d_l(p) > 0\} \qquad (9.6)$$

We note that this formulation creates the confidence about the lidar points, as an output. Meanwhile, to increase the robustness of the model, it might be nice to provide a confidence estimate like this as an input to the model. This would require a two stage model, such as the one proposed in [32]. Additionally, while we calculate the weight for the entire image, due to the fully convolutional architecture, we only constrain and take it into consideration at the distinct set of points where we have lidar points. Everywhere else it is completely unconstrained and has thus no meaning.

Photometric Loss

A popular loss we included has been proposed by [44, 26], which was already presented in Section 9.3.2. The authors in [26] remark "In the self-supervised framework, the training process is more iterative [...], it takes many more iterations for the predictions to converge to the correct value". We investigated the relationship between the prediction error and the photometric loss. We observe a very large variance in the loss, even if we provide ground truth labels as input.

We can improve upon this, by not considering only one nearby frame, but rather an average of the photometric loss over multiple nearby frames, both backwards and forwards in time. We take a total of 6 frames, and calculate the photometric error as the average between those frames. If the PnP solver does not find a solution for the pose between two frames, instead of replacing it with an identity mapping, as is done in [26], we completely drop it.

Smoothness Loss

Depth is, for very large regions, a continuous measure: Compared to the image size, there are very few regions constituting hard edges. In a first assumption, the detected scene can be approximated as a set of planes which have a random orientation in space. The depth loss formulated above produces high quality mappings for the points at which we have lidar information available, but will falter for all other points. We can thus employ a smoothness constraint, so that the model finds a function which still fulfills the identity mapping we define above, while giving a smooth estimate for the depth change between those points.

In [26], this constraint is defined as

$$\mathcal{L}_{\text{smooth}} = \sum_p \left(\left\| \nabla^2 (\hat{d}(p)) \right\|_2 \right) \tag{9.7}$$

In this case the nabla operator is approximated using first order differences, so that ∇^2 can be expressed at $2\hat{d}(p) - \hat{d}(p-1) - \hat{d}(p+1)$, summed in both directions. This smoothness constraint has an effective field of view 3. However, smoothness is not such a localized property but often stretches over large parts of the image, for example for a wall. We observed that the model will learn single points of correct predictions, due to the identity loss, surrounded by a region of pixels with a constant gradient, going down to 0 distance again.

Therefore, we propose to introduce a multi-scale smoothness constraint

$$\mathcal{L}_{\text{smooth}} = \sum_{s \in S} \sum_p \left(\left\| \nabla^2 (\hat{d}^{(s)}(p)) \right\|_2 \right) \tag{9.8}$$

Multi-scale losses are a common way to enforce size independence in a variety of tasks, such as object detection or segmentation [23, 26, 42]. We use a total of 5 different sizes, meaning that the predicted depth is downsampled up to a size of only 22 by 76 pixels. In such a size, a lot of objects are less than a few pixels in size, violating our assumption that the number of pixels, which correspond to a hard edge, is far smaller than the total number of pixels in the image.

Therefore, we want to find a formulation, which constraints the smoothness at points far from object borders, also over larger distances, while allowing for large changes in depth at object edges. To obtain a notion of object, we use semantic segmentation: Using a state-of-the-art model [43], which is trained on the Cityscapes dataset, we generate semantic segmentation masks for all images in our dataset. The masks produced are divided into 19 classes. After generating the probabilities for all pixels in the image, we apply NMS (non-maximum suppression) to obtain the most likely class and store this information. Using the probabilities to generate richer auxiliary information was not done, due to the large memory requirements of storing 19 floating point number per pixel. Due to the hard decision, we can reduce the memory for a single pixel to 1 byte.

The borders of objects can then be calculated as a binary decision. A given pixel lies on an edge, if it is not equal to all its neighbors. We then apply binary dilation to grow our edge regions to take non perfections in the masks and alignment into consideration. When we calculate our multiscale images, we use a pessimistic approach to determine the downsampled edge regions: If any of the four pixels which make up the downsampled pixels was on an edge, the pixel is considered to be on an edge in the downsampled version.

Figure 9.8 shows the results of such the mask generation. We can see how the masks grow for greater downsampling factors (bottom left to bottom right). Due to the square nature, we introduce some aliasing. On the right hand side of the picture, best seen in the top right overlay, we have a number of wrong speckles. While we could filter the obtained segmentation masks to remove such small regions, we do not have to consider them. We apply our smoothness constraint on the inverse of the maps shown: Only outside of the edge regions, we want to enforce smoothness. Thus, false positive edges are of no concern to us. They reduce the smoothness cost at that location, which only decreases the gradient on the loss we want, but not the direction.

Figure 9.8: Visualization of the masking applied for the multiscale smoothness loss. Top left is the input image, top right with dilated semantic borders. In the bottom part, the mask is shown, downsampled by a factor of 2 and 8, respectively.

Formally, we can write our loss function finally as

$$\mathcal{L}_{\text{smooth}} = \sum_{s \in S} \sum_{p} \left(\left\| \nabla^2 (\hat{d}^{(s)}(p)) \right\|_2 \right) \forall \{ p \in P \big| m^{(s)} = 0 \} \tag{9.9}$$

with $m^{(s)}$ being the downsampled mask.

9.5 Results

In the following section, we will present some numeric results for the approach we are proposing. Our investigations show that the proposed approach outperforms the baseline model. We introduce a measure to determine if a point is correctly identified as an outlier and show that our confidence prediction can detect them to some degree.

Unfortunately, due to compute constraints, we did not perform an ablation study of the different design decisions we made, though this is certainly an interesting future topic to discuss. The convergence of the baseline and our proposed model is not completely stable in the sense that the results vary run over run. The proper validation way would be to train the model multiple times and report results in the average, worst and best case. However, due to the small difference in performance, we believe that the results we show stand by themselves, as there is a fairly significant margin.

9.5.1 Confidence Estimation

For a given pair of sparse and semi dense depth images, we want to qualify the quality of our confidence estimation. We would want the confidence to be low, if a point in the sparse depth map is false, while it should be close to 1 if the point aligns well with the ground truth.

We can thus divide the set of points into three distinct classes: Positive points, where we have a depth measure available both in the sparse and semi dense image, and the error between them is small. As cut-off for what we consider small, we chose an squared error of 0.5 m. Negative points, where we have ground truth and sparse depth measures available, but the error is greater than 0.5 m. And finally, we have points, where we are not sure, if they are correct, but they are more likely to be false. For these points, we have no ground truth available, but only have sparse depth measurements.

Our results indeed show, that the model has successfully learnt to distinguish between positive and negative points. Ignoring the unsure points, the problem turns into a binary classification problem. We can then show the results in a ROC curve, which is shown in Fig. 9.9. Note that, different to most literature, we show the ROC for the negatives instead of the positives. We have a lot of true positives, but only few true negatives. We thus are interested in filtering out as many true negatives, even at the cost of removing a lot of true positives as well.

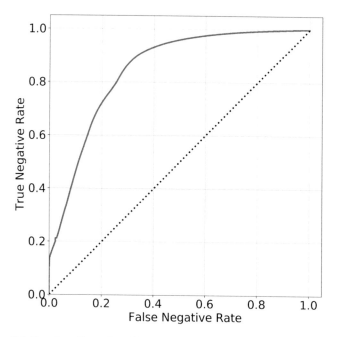

Figure 9.9: Receiver Operating Characteristics curve of the confidence estimation.

If we choose 0.99 as confidence threshold, we obtain the confusion matrix in Table 9.2. The last row and last column correspond to the total number of predictions, or truth values respectively. With such a threshold, we remove 80% of the outliers (true negatives), while removing only 25% of the inliers (false negatives).

Table 9.2: The confusion matrix for a threshold of 0.99

		Confidence		Total
		<0.99	>0.99	
Truth	Negative	1565903	391476	1957379
	Positive	18272485	53200085	71472570
Total		19838388	53591561	

9.5.2 Depth Estimation

Table 9.3 shows an overview of the results we obtained. Unfortunately, we were unable to reproduce the results presented in [26], which appear to be state-of-the-art for self-supervised depth estimation using deep learning on the KITTI dataset. Our own results, using the code of the original authors, are approximately 90 mm worse in terms of RMSE.

Table 9.3: Comparison between various methods for unsupervised depth estimation on the KITTI dataset. We could not reproduce the results as reported in [26], so we show the results obtained using a smaller network and their method, as well as the results they reported (first two rows)

Method	RMSE	MAE	SiLog	Rel SE	IRMSE	IMAE	Abs Rel	$\delta 1$
[26] Own	1470.10	372.80	5.47	7.55e-3	4.67	9.89	0.023	0.991
[26]	1384.85	358.92	-	-	4.32	1.60	-	-
[19]	1312.86	287.02	5.00	5.20e-3	3.88	1.22	0.015	0.993
DL Ours	1174.35	319.50	4.41	4.4e-3	3.72	1.48	0.015	0.994
[19]+Conf	723.64	180.98	4.71	4.3e-3	4.72	1.50	0.016	0.994
Average	1180.42	290.39	4.41	4.2e-3	3.61	1.29	0.023	0.994

The second baseline we compare against are the result obtained by [19], which is using traditional image processing techniques to obtain a depth map. We compare these results against the performance of our own experiments:

DL Ours The deep learning model we trained in an unsupervised fashion, as discussed above, with the loss terms we introduced.

[19]+Conf The traditional computer vision approach, but with all the low confidence values masked out. Low confidence in this case means a confidence less than 0.99. Note that this method does not generate completely dense maps.

Average Taking the elementwise average, where available, between the self supervised and the cv model.

We can see that our deep learning model outperforms state-of-the-art unsupervised methods by a significant margin. While it is very impressive to see the low RMSE error for the filtered traditional cv approach, reaching an RMSE of 723.64 mm, better than the state-of-the-art supervised methods on the KITTI dataset (736 mm by [37]). However, as the results are not dense, there are holes in the predictions we have to fill. These areas are especially difficult, thus making the fused models, on average, perform not as well. In Fig. 9.10, we show the percentage of predicted pixels, relative to the number of pixels for which ground truth is available, as a function of the ground truth distance. It shows how the baseline approach does not predict every single pixel, with a decreasing percentage for larger distances, as well as the significantly reduced number of pixels predicted by the baseline, if it is fed with the filtered sparse depth. Pretty much no predictions are made for points further than 30 m away. In contrast, the deep learning models always produce dense predictions. This is an indicator for one of the downsides of using a traditional computer vision approach: The entire method is finetuned to the dataset, and adaptation to a changed distribution, as is the case for the filtered point clouds, requires manual tuning of parameters such as kernel size.

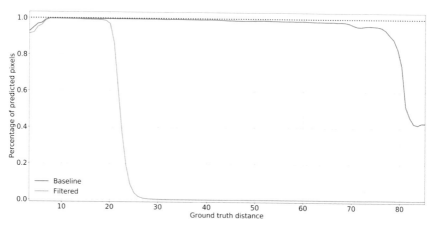

Figure 9.10: Percentage of ground truth pixels, as a function of distance, for which the baseline [19] and the filtered approach ([19]+Conf in Table 9.3) produce predictions.

We show the results for the deep learning, minimum and average fusion model for pixels, which are available in the confidence filtered cv approach in Table 9.4. They show that the better performance is mainly due to the predicted regions instead of an actually better performing model. For that subset of pixels, the averaging between the deep learning and the traditional cv model performs the best in terms of RMSE.

Table 9.4: Comparison of our methods, on a subset of pixels. The metrics were only calculated on the pixels which were predicted by the "[19]+Conf" approach

Method	DL Ours	[19]+Conf	Minimum	Average
RMSE	629.4984	723.64	656.5946	587.3359
MAE	299.9958	180.98	165.6115	209.7372
SiLog	3.6829	4.712	3.8128	3.7972
Rel SE	3.4e-3	4.3e-3	1.9e-3	2.9e-3
IRMSE	4.801	4.72	3.5452	4.4516
IMAE	2.8357	1.50	1.3356	2.004
Abs Rel	0.02791	0.016	0.01364	0.01962
δ_1	0.9956	0.994	0.9962	0.9944
δ_2	0.9983	0.9967	0.9984	0.998
δ_3	0.9992	0.9984	0.9992	0.9993

Looking at the fused models, we see that averaging does not improves the performance in terms of RMSE relative to the deep learning's output, but in most other metrics. However, in a safety critical environment, such as object detection for autonomous driving, decreasing the nominal performance (by increasing the RMSE), might be a worthwhile trade-off: It is better to systematically underestimate the distance to possible obstacles, thus using pixelwise minimum as fusion approach might be a more valid idea than the raw performance characteristics might indicate (not shown).

For the following comparisons, we show only the baseline from [19], as we cannot reproduce the results from [26] due to hardware constraints. If we were to show the results based on that work with the lower complexity model (first column in Table 9.3), we believe that the conclusions would be more difficult to draw, as the model performs significantly worse than the reported values of the original authors.

We show a visual comparison in Fig. 9.11. The first column corresponds to (in order), the input RGB image, the sparse depth cloud and the ground truth data. The second column shows the [19] baseline, our deep learning model and finally the proposed averaged model. The outputs contain different types of artefacts, depending on the approach. The baseline output shows an error made at the top of the vehicle, where it significantly extends above the actual location. We also see that there is no results for large distances, indeed, the maximum predicted distance by the baseline is in the range of 40 m. On the right hand side, one can also observe the blockier output of the baseline, due to the limited perceptive field of the filters.

Using the smoothness and photometric constraints, the proposed model is able to learn some depth prediction at points above the field of view of the lidar: The speed limit sign on the right hand side is assigned a realistic depth. The smoothness constraint has the downside of penalizing large jumps in depth. Thus there is a small halo around the leading vehicle where the depth rapidly increases, but the edge is not as sharp as it should be. It thus may be appropriate to apply a non-linearity to the second-order-differential. Instead of the formulation in Equation (9.9), applying a non-linearity to the L^2 norm, which clamps the loss

Figure 9.11: Visualization of the results we obtained. Left: Inputs and ground truth. Right: Baseline, deep learning and averaged model. Best viewed in color.

at some maximum second-order derivative, could improve upon this: The loss would be a function of the smoothness, but would saturate at some point. From that point, the depth could increase in steepness at no additional loss.

In summary, we show that the proposed deep learning model outperforms other state-of-the-art methods for unsupervised depth completion. The performance is further improved by employing model fusion using averaging of the outputs. Traditional computer vision approaches can, for some areas, still compete with deep learning methods. However, we also presented approaches superior to these models, and show that removal of detected outliers improves the performance of the model.

9.6 Conclusions and Future Work

In this paper, we have shown that, using unsupervised learning, we can learn to estimate depth from a single RGB image and a synchronized lidar frame and differentiate faulty measures from real ones.

As computer vision is a very hot topic due to the popularity of deep learning, since the start of this work, the state-of-the-art has improved considerably in the depth estimation task. Supervised approaches start to reach an RMSE of 700 mm. Using their architectural tricks, such as guided convolution [37], performance improvements using self supervision could also be obtained. Recent work also showed great promise of late feature fusion: Instead of concatenating the image and lidar at the input of the model, the high dimensional feature space in the UNet architecture can be used, or a mix where every feature level is concatenated for all input modalities. While supervised approaches outperform unsupervised setups such as the one we proposed, we believe that the high complexity of data gathering, cleaning and annotating can be avoided with well designed loss functions are experimental setups for unsupervised learning.

Finally, as every difference between the dataset and the final environment can drastically decrease the performance of deep learning models, we believe that gathering datasets which match the final setup as close as possible is a worthwhile endeavor. If self-supervised approaches are employed, the manual labor required is relatively small. A more comprehensive setup would allow the data to be more useful. As it is easier to discard data after the fact, a well-equipped measurement vehicle, providing power to the PC, a large sensor array, and fast and large amounts of storage, enables more comprehensive analysis of driving scenarios.

References

1. AutonomouStuff. 2019. Velodyne Lidar Alpha Puck Datasheet.
2. Baidu. 2017-2020. Apollo github repository. https://github.com/ApolloAuto/apollo.

3. Baidu. Apollo open platform. https://apollo.auto.
4. Bay, H., A. Ess, T. Tuytelaars and L. Van Gool. 2008. Speeded-up robust features (surf). *Comput. Vis. Image Underst.*, 110(3), 346–359.
5. Bradski, G. 2000. The OpenCV Library. *Dr. Dobb's Journal of Software Tools.*
6. Casser, V., S. Pirk, R. Mahjourian, and A. Angelova. 2019. Depth prediction without the sensors: Leveraging structure for unsupervised learning from monocular videos. *In*: Thirty-Third AAAI Conference on Artificial Intelligence (AAAI-19), 2019.
7. Chen, L-C., Y. Zhu, G. Papandreou, F. Schroff and H. Adam. 2018. Encoder-decoder with atrous separable convolution for semantic image segmentation. *In*: ECCV.
8. D. Ciresan, A. Giusti, L. M. Gambardella and J. Schmidhuber. 2012. Deep neural networks segment neuronal membranes in electron microscopy images. pp. 2843–2851. *In*: F. Pereira, C.J.C. Burges, L. Bottou and K.Q. Weinberger (ed.). Advances in Neural Information Processing Systems 25. Curran Associates, Inc.
9. Cordts, M., M. Omran, S. Ramos, T. Rehfeld, M. Enzweiler, R. Benenson, U. Franke, S. Roth and B. Schiele. 2016. The cityscapes dataset for semantic urban scene understanding. *In*: Proc. of the IEEE Conference on Computer Vision and Pattern Recognition (CVPR), 2016.
10. Dietterich, Thomas G. 2000. Ensemble methods in machine learning. pp. 1–15. *In*: Proceedings of the First International Workshop on Multiple Classifier Systems. MCS '00. Springer-Verlag. London, UK.
11. Galler, B.A. and M.J. Fisher. 1964. An improved equivalence algorithm. *Commun. ACM*, 7(5), 301–303.
12. Geiger, A., P. Lenz, C. Stiller and R. Urtasun. 2013. Vision meets Robotics: The KITTI Dataset. *International Journal of Robotics Research (IJRR).*
13. Geiger, A., P. Lenz and R. Urtasun. 2012. Are we ready for autonomous driving? The KITTI vision benchmark suite. *In*: Conference on Computer Vision and Pattern Recognition (CVPR).
14. Girshick, R., J. Donahue, T. Darrell and J. Malik. 2014. Rich feature hierarchies for accurate object detection and semantic segmentation. pp. 580–587. *In*: Proceedings of the 2014 IEEE Conference on Computer Vision and Pattern Recognition, CVPR '14. IEEE Computer Society. Washington, DC, USA.
15. SAE International. 2017. Automated driving levels of driving automation are defined in new sae international standard j3016.
16. Ioffe S. and C. Szegedy. 2015. Batch normalization: Accelerating deep network training by reducing internal covariate shift. 448–456.
17. Isola, P., J-Y. Zhu, T. Zhou and A.A. Efros. 2016. Image-to-image translation with conditional adversarial networks. pp. 5967–5976. *In*: 2017 IEEE Conference on Computer Vision and Pattern Recognition (CVPR).
18. Norman, P. Jouppi et al. 2017. In-datacenter performance analysis of a tensor processing unit. pp. 1–12. *In*: Proceedings of the 44th Annual International Symposium on Computer Architecture, ISCA '17. ACM. New York, NY, USA.
19. Ku, J., A. Harakeh and S.L. Waslander. 2018. In defense of classical image processing: Fast depth completion on the CPU. pp. 16–22. *In*: 2018 15th Conference on Computer and Robot Vision (CRV). IEEE.
20. Ku, J., M. Mozifian, J. Lee, A. Harakeh and S. Waslander. 2018. Joint 3d proposal generation and object detection from view aggregation. *IROS.*
21. Kuhn, H.W. and B. Yaw. 1955. The hungarian method for the assignment problem. *Naval Res. Logist. Quart.*, 83–97.

22. Lambert, F. 2019. Tesla leaks info about new self-driving computer in latest software update. https://electrek.co/2019/01/04/ tesla-leaks-hardware-3-self-driving-computer, 2019.

23. Liu, W., D. Anguelov, D. Erhan, C. Szegedy, S. Reed, C.-Y. Fu and A. Berg. 2016. SSD: Single Shot MultiBox Detector. 9905, 21–37.

24. Lowe, D.G. 2004. Distinctive image features from scale-invariant keypoints. *Int. J. Comput. Vision*, 60(2), 91–110.

25. Lynen, S., M. Achtelik, S. Weiss, M. Chli and R. Siegwart. 2013. A robust and modular multi-sensor fusion approach applied to mav navigation. *In*: Proc. of the IEEE/RSJ Conference on Intelligent Robots and Systems (IROS).

26. Ma, F., G. Venturelli Cavalheiro, and S. Karaman. 2019. Self-supervised sparse-to-dense: Self-supervised depth completion from lidar and monocular camera. ICRA.

27. Ma, F. and S. Karaman. 2018. Sparse-to-dense: Depth prediction from sparse depth samples and a single image. ICRA.

28. Maas, A.L., A.Y. Hannun and A.Y. Ng. 2013. Rectifier nonlinearities improve neural network acoustic models. ICML Workshop on Deep Learning for Audio, Speech and Language Processing.

29. Mahjourian, R., M. Wicke and A. Angelova. 2018. Unsupervised learning of depth and ego-motion from monocular video using 3d geometric constraints. CVPR.

30. Menze M. and A. Geiger. 2015. Object scene flow for autonomous vehicles. *In*: Conference on Computer Vision and Pattern Recognition (CVPR).

31. Nvidia. Cuda toolkit. https://developer.nvidia.com/cuda-toolkit.

32. Qiu, J., Z. Cui, Y. Zhang, X. Zhang, S. Liu, B. Zeng and M. Pollefeys. 2019. Deeplidar: Deep surface normal guided depth prediction for outdoor scene from sparse lidar data and single color image. pp. 3313–3322. *In*: Proceedings of the IEEE Conference on Computer Vision and Pattern Recognition.

33. Redmon, J., S.K. Divvala, R.B. Girshick and A. Farhadi. 2015. You only look once: Unified, real-time object detection. pp. 779–788. *In*: 2016 IEEE Conference on Computer Vision and Pattern Recognition (CVPR).

34. Ronneberger, O., P. Fischer and T. Brox. 2015. U-net: Convolutional networks for biomedical image segmentation. *Medical Image Computing and Computer-Assisted Intervention (MICCAI)*, volume 9351 of *LNCS*, 234–241 (available on arXiv:1505.04597 [cs.CV]).

35. Shelhamer, E., J. Long and T. Darrell. 2017. Fully convolutional networks for semantic segmentation. *IEEE Trans. Pattern Anal. Mach. Intell.*, 39(4), 640–651.

36. Su, Z., M. Ye, G. Zhang, L. Dai and J. Sheng. 2019. Cascade feature aggregation for human pose estimation.

37. Tang, J., F-P. Tian, W. Feng, J. Li and P. Tan. 2019. Learning guided convolutional network for depth completion.

38. Uhrig, J., N. Schneider, L. Schneider, U. Franke, T. Brox and A. Geiger. 2017. Sparsity invariant CNNS. 2017 International Conference on 3D Vision (3DV), pp. 11–20.

39. Wu, Y. and K. He. 2019. Group normalization. *International Journal of Computer Vision*.

40. Wu, Z., C. Shen and A. van den Hengel. 2016. Wider or deeper: Revisiting the resnet model for visual recognition. *ArXiv*, abs/1611.10080.

41. Xu, B., N. Wang, T. Chen and M. Li. 2015. Empirical Evaluation of Rectified Activations in Convolutional Network.

42. Xue, Y., T. Xu, H. Zhang, L.R. Long and X. Huang. 2017. SegAN: Adversarial Network with Multi-scale $L1$ Loss for Medical Image Segmentation. *CoRR*, abs/1706.01805.
43. Sapra, K. and Y. Zhu. 2019. Improving semantic segmentation via video propagation and label relaxation. *In*: IEEE Conference on Computer Vision and Pattern Recognition (CVPR).
44. Zhou, T., M. Brown, N. Snavely and D.G. Lowe. 2017. Unsupervised learning of depth and ego-motion from video. CVPR.

Artificial Intelligence Based on Modular Reinforcement Learning and Unsupervised Learning

Shawan Taha Mohammed[*], Angelo Mihaltan, Kemal Acar, Hendrik Laux and Gerd Ascheid

RWTH Aachen University

10.1 Introduction

10.1.1 Motivation

Reinforcement learning (RL) has achieved great development in the past years. Besides the development of many algorithms like DQN [18], DDPG [14], A3C [17] or PPO [23] many deep learning (DL) tricks have been applied in this area. The progress observed on simple multilayer perceptrons (MLP) regarding convolutional neural networks (CNN) and recurrent neural networks (RNN) can be an accurate example in this manner. As well as using modular architectures for the neural network or using unsupervised learning practices and RL in combination. The last two methods are discussed in this chapter.

RL opens a great opportunity for the artificial intelligence (AI) world. This opportunity rises due to the fact that, with RL, rather than expensive, human labeled data, a simple feedback signal is used and with this signal complex strategies can be derived. This feedback signal is called reward function in the context of RL. An agent is rewarded if a certain action was good. The agent can either be "suspended" in reality or in a simulation and it should try to learn a certain task. From a simple reward and punishment approach, the agent can develop complex strategies using RL algorithms and gradient descent optimization [GD 2019]. The ambition of RL scientists is to have an agent that can solve complex tasks from

---

*Corresponding author: Shawan.mohammed@ice.rwth-aachen.de

even the simplest reward function. But there is still a long way to go. Currently, it is still very difficult to learn small tasks if the reward function is not "informative" enough. Let's assume we have a simple reconstruction task as in the Fig. 10.1. You use a neural network to create output that is identical to the original input $(X - \hat{X})$. This is a simple task for a neural network, since it only needs to pass input. To train such a neural network, a gradient descent optimization is used. As with any optimization, a loss function is needed to measure the error. As shown in Fig. 10.1, an unsupervised learning loss ($Loss_{USL}$) or a reinforcement learning loss ($Loss_{RL}$) can be used. It is obvious that $Loss_{USL}$ provides much more information than $Loss_{RL}$ by comparing the prediction and the desired output. $Loss_{RL}$ provides a sparse reward if the prediction is equal to the desired output, the loss function is one, otherwise it is zero. This simple task shows the main challenge in Deep RL. Through RL, millions of parameters must be optimized starting from a simple 35 reward function. Obviously, the optimization cannot work well if the feedback signal $Loss_{RL}$ itself does not contain much information except a "bare good or bad". Moreover, even the gradient descent method itself is a relatively noisy optimization method.

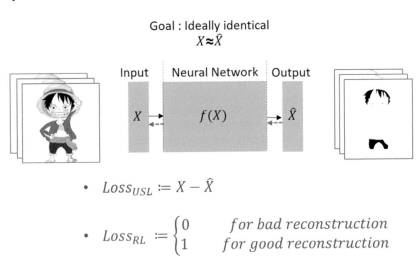

$$Loss_{USL} := X - \hat{X}$$

$$Loss_{RL} := \begin{cases} 0 & for\ bad\ reconstruction \\ 1 & for\ good\ reconstruction \end{cases}$$

Figure 10.1: Example task to show the difference between unsupervised and reinforcement learning loss.

10.1.2 Goal

In summary, RL is a comprehensive way to develop a general AI but also a more difficult variant from the optimization point of view. The deeper the networks become, the more difficult it is to optimize the millions of parameters by just sparse rewards. Reducing the parameters that are optimized via RL can make the problem much easier. In this work, we address exactly this problem by optimizing only a part of the parameters via RL, which can only be optimized via RL. A simple and relatively obvious division of a neural network is to divide it into a feature extraction (FE) and a logic part. The FE is only responsible for the mere splitting of inputs into features and can be trained with supervised learning or unsupervised learning, in many machine learning frameworks it is also called perception module like in the Apollo framework [2]. The logic layer, in our case the actor network, is the only network that is trained via RL. Thus, we combine unsupervised learning and RL. To accomplish it, we use a variational autoencoder and connect it to the actor-critic architecture of RL.

10.1.3 Related Work

The idea to combine variational autoencoder (VAE) and RL is not new. In the paper [6] a VAE in combination with RL is used. They call their framework world models and they use a VAE and unsupervised learning to train an encoder which provides a compressed spatial and temporal representation of the environment to the RL agent. Shortly thereafter, [8] have taken a similar approach in their work on "Deep Variational Reinforcement Learning for POMDPs". Also here a VAE was trained and the output of the encoder was used as input for an actor-critic RL framework. Furthermore, the thesis also showed that a combined training of the encoder through unsupervised learning and RL increases the performance.

10.2 Machine Learning

Machine Learning (ML) is a sub-area of artificial intelligence. It is a data-driven method and aims to recognize patterns in data. For this purpose, mathematical models are trained using algorithms and data. This section provides theoretical background on machine learning. This is very important because in this thesis a framework is presented, which consists of different neural networks and trained by different optimization algorithms. In order not to exceed the scope of this chapter we limit ourselves to the division of ML into the general three classes:

1. Supervised learning (SL)
2. Unsupervised learning (USL)
3. Reinforcement learning (RL)

We give an overview of the three different types of ML classes. Afterwards, we go further into deep learning and neural network architecture.

10.2.1 Supervised-Unsupervised-Reinforcement Learning

Supervised learning is one of the most popular ML classes. It uses labeled data. This means that you know exactly what output you can expect from the model when certain data is given as input. The training takes place as follows:

Data is transferred to the model, then a prediction is made by it. The obtained output of the model is compared with the expected output, known as the ground truth. The difference between the prediction of the model and the ground truth is a cost with which the model is afterwards trained. It is expected that the trained model generalizes enough. So that it can accurately predict unknown data without having seen it during training. In general, there are two problems for which supervised learning is used. First, a classification problem. A model is supposed to assign certain inputs to specific classes or discrete values. Second, a regression problem. Here you try to model the relationship between an input and a continuous output. Although supervised learning is a favorable and reliable way of ML, it has a major drawback. Namely, generating databases is a very expensive task, because people have to label them by hand. This leads us to another learning class: unsupervised learning. This group of learning algorithms use unlabeled data. During the training, it is tried to figure out the relations of elements within a database. Equality or anomalies are investigated and the elements are classified accordingly. Another application of this class is the compression of information using autoencoders. Also known as dimensionality reduction. It is a very helpful tool for ML users and will be discussed later in this chapter. The third large group of learning algorithms is RL. Like unsupervised learning, there is no need for labelled data. In this variant, the data itself is generated by interaction with an environment. This can be the reality but also a simulation or a game. The environment receives an action from the model and outputs a reward and a resulting state. The goal of RL is to maximize this reward.

10.2.2 Deep Learning

In the course of the development in the last decade, it was impressively shown that all ML algorithms mentioned above gained performance when used in combination with deep neural networks. This endeavour falls under the term deep learning. Deep learning is a machine learning class that uses artificial neural networks (ANN). Based on the Rosenblatt Perceptron from 1963 [15], there is wide range of ANN structures. The gradient descent method is mainly used as an optimization method. Since the development of AlexNet [10], constantly better results have been achieved in various areas. For example from the field of Computer Vision, which deals with perception. Starting from raw 2D images, patterns are deduced to clearly identify and locate people or animals in the image. Also in other areas, language modeling like in the work of Radford [20] or audio recognition like in the work of Ramon [21] or Geitgey [4] deep learning approaches are becoming increasingly important. In order not to exceed the scope of this chapter we refer to the book "Deep Learning" [5]. This book deals with deep learning, ANNs, optimizers and other related topics in detail and explains them very adequately.

10.2.3 Advanced Deep Neural Network Architectures

Throughout the development of ML and deep learning, some network architectures have become established. The best known example is CNNs developed [12], they are used for image processing. Furthermore, RNNs are one of the most important network types. They are used to walk time correlated sequences of data. Autoencoder, variational autoencoder and for RL typical actor-critic architecture will be discussed in more detail in the following.

10.2.3.1 Autoencoder

Autoencoders (AE) has a long tradition in machine learning, especially in the area of unsupervised learning methods. The aim is to reduce a high dimensional space into a lower dimensional latent representation.

Figure 10.2 depicts schematically the structure of an AE. The input X is compressed via an encoder into a latent space Z. Then follows a reconstruction from Z to $\hat{X}$ via the decoder into the original dimension. The reduction and subsequent reconstruction from a bottleneck is implemented in deep learning via neural networks [3]. AEs are often used to compress or encrypt information of any kind by using latent space for information transfer or storage. An instance for distribution over the latent space for a trained encoder is given in Fig. 10.3. The AE has been trained on the MNIST data set [13]. It appears that for each digit the encoder learns an illustration of clusters that have irregular shapes with sparse regions.

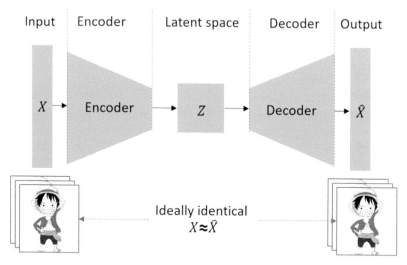

Figure 10.2: Schematically structure of an autoencoder.

10.2.3.2 Variational Autoencoder

The vanilla AE has a sparseness problem. It tends to form a latent space where the latents of different classes and features are widely separated, see Fig. 10.3. This is because the AE only increases the probability of correctly reconstructing the input

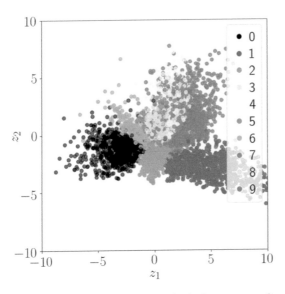

Figure 10.3: Distribution of the latents in the latent space of an AE.

data from a statistical point of view. The overall structure learns to be very secure in the encoding and decoding procedure and is not able to deal with uncertainties. However, if the decoder receives a latent from a sparse region, the decoder would not reconstruct anything useful.

An extension to autoencoders is provided by the work [9] via variational autoencoder (VAE). The VAE shown in Fig. 10.4 is similar to the AE structure shown in Fig. 10.2. It also contains an encoder component that learns the dimensional reduction and a decoder component that aims to reconstruct the input information. However, instead of mapping the input data set into the latent space, the encoder learns how to map the input into a certain distribution. Afterwards, the distribution is sampled randomly and then transferred into a latent vector. This change in the bottleneck leads to a significant change in learning behavior. An Elbo loss function is used to influence the distribution in the bottleneck. The aim is to force gaussian-like distribution $N(\mu, \sigma)$ to get an advantage in reconstruction. A gaussian distribution is an established method in ML and is often used. It is acknowledged how to reconstruct, cluster or sample from this distribution. The VAE proposes a solution for efficient posterior inference, that is needed to get control over the latent space for the purpose of efficient codings. The derivations and some other theoretical stuff can be looked up in the papers "Structured Stochastic Variational Inference" [7] "Auto-Encoding Variational Bayes" [7] and also from an article "Understanding Variational Autoencoders (VAEs)" [22].

$$\mathcal{L} = C\left\|X - \hat{X}\right\|^2 - D_{KL}\left(N(\mu,\sigma)\right)\left\|N(0, I)\right. \tag{10.1}$$

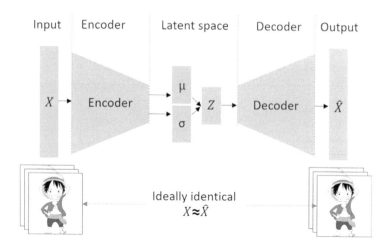

Figure 10.4: Schematically structure of variational autoencoder.

Eq. 10.1 describe the VAE loss. The first term of the loss is a so-called reconstruction loss. This is the same loss that is used in a normal AE. It ensures that the input X and output $\hat{X}$ are identical. The second term of the loss is the Kullback-Leibler divergence and it acts as a regularizer. This forces the encoder to deliver a gaussian-like output with zero mean and one variance N(0, I). Figure 10.5 illustrates the effect of the VAE loss. The image was created under the same conditions as Fig. 10.3. In contrast to Fig. 10.3, however, the latent space is used optimally and the latents are located around the zero point. The influences of the individual terms in relation of latency can be seen in Fig. 10.5.

10.2.3.3 Actor Critic Scheme

In the field of deep RL there are many algorithms that can be roughly divided into two categories: Policy gradient based and value function based methods.

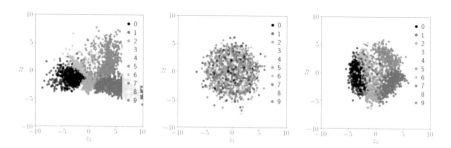

Figure 10.5: Distribution of latents in the latent space of a VAE under the influence of different loss functions. The loss function in the left image consists only of a reconstitution loss $\|X - \hat{X}\|_2$ and in the middle image only of a Kullback-Leibler divergence loss $D_{KL}(N(\mu; \sigma)\|N(0; I))$. The right picture is the combination of both and represents the normal VAE loss, as depicted in Eq. 10.1.

However, the algorithms of both groups have advantages and disadvantages. One architecture that tries to combine the advantages of both groups is the actor-critic scheme. It works with an algorithm which is both policy gradient based and value function based. Figure 10.6 shows the structure of such an architecture. First we have an environment from which we receive the states and rewards. These will be passed as inputs to two neural networks, the critic and the actor. The critic is value function based and tries to assign a value to each state. The actor is policy gradient based and tries to output the best action that maximizes the cumulative reward. Thereby it is important to know that with the output of the critic the actor is trained. One of these RL actor-critic algorithms is the "Proximal Policy Optimization Algorithms" (PPO) [23]. It is currently one of the state of the art algorithms and is used in this thesis to train the actor.

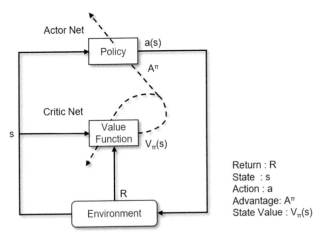

Figure 10.6: Actor-critic architecture.

10.3 Methodology

In section 10.2 the different classes of ML were discussed as well as AE and VAE. Unfortunately, the topics deep RL, CNN, RNN and especially PPO as the RL algorithm favored by us, were not further explained. Nevertheless this is necessary to understand the proposed methodology in depth. Since we are limited in the context of a book chapter, we refer the interested reader to the work [16]. With this chapter we would like to motivate the reader on the modular RL and the use of several ML classes to build a general artificial intelligence (GAE). Therefore it is very important to understand the differences between RL, USL and SL and explain the structure of the machine learning section. This allows for the creation of a modular architecture consisting of a perception and a logic module and combine their individual advantages.

Figure 10.7 illustrates the structure of the proposed framework. We obtain inputs from the environment. These are passed as input to the perception module

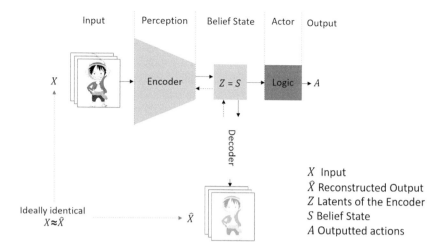

Figure 10.7: Schematic structure of the modular framework with a separate perception and logic module.

during execution. The perception module maps the inputs into a latent vector. In the work [8] this is considered as a belief state. The latent vector is passed as input to the actor-critic framework. Then an action according to the policy of the actor is passed to the environment. In the training case, the framework behaves differently than in the execution. For example, there is no backpropagation path from the action of the actor back to the perception module. Instead, only the logic part is trained only via RL. The perception module is trained in an unsupervised way. This works because it is nothing more than an encoder during training and forms a VAE in combination with a decoder.

10.4 Simulation

The framework developed in section 10.3 is applied to the simulation CarRacing-v0, which is a standardized simulation of OpenAI [19]. It is free and easy to install. It will be installed via the terminal and can then be imported into python as a library. The simulation offers ideal conditions to evaluate RL algorithms. There are uniform inputs and outputs and the reward function is given. This makes a comparison with other algorithms very transparent.

10.4.1 Output States

The CarRacing environment provides states by images, see Fig. 10.8. They are 96×96 large pictures in RGB format.

10.4.2 Input States

As input the environment accepts three continuous actions.

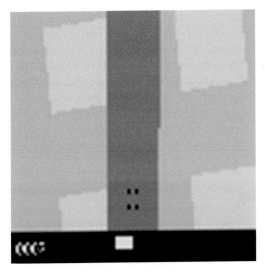

Figure 10.8: A typical CarRacing input image for an RL agent.

1. Steering $\{a^s \in \mathbb{R} | -1 \leq a^s \leq 1\}$
2. Gas $\{a^g \in \mathbb{R} | 0 \leq a^g \leq 1\}$
3. Brakes $\{a^b \in \mathbb{R} | 0 \leq a^b \leq 1\}$

10.4.3 Reward

As mentioned above, the environment also provides rewards. This represents a measure of how good the agent's actions were in the environment. In the case of CarRacing, the agent receives a penalty of -0.1 for each action he takes, and is only rewarded if he picks up a tile on the street. These tiles are randomly distributed on the road and the agent receives a reward if it drives over one.

10.4.4. Difficulties and Challenges

At first glance CarRacing seems to be an easy to solve simulation or task. But from an optimisation point of view, CarRacing faces ML algorithms of all kinds with three major difficulties:

1. The first difficulty of the CarRacing environment is due to the fact that it works with images. Images are tensors that contain brightness values. These must be decomposed into information by a feature extraction part from which logical conclusions can then be derived. It is similar to the human optical nervous system. The retina absorbs light. This is passed on to a perception module until patterns and complex structures can be recognized.
2. The next difficulty is the environment works with continuous actions. Unlike discrete environments, there are theoretically infinite possibilities of actions that the agent can execute.

3. The third and last point is that the reward function is difficult defined for an optimization problem like this. It must be optimized over two terms, namely collect tiles and make few actions or steps in the environment. Furthermore the agent only gets a reward when it passes over a tile. The surface of such a tile is very small compared to the arena the agent is in. From the perspective of an agent there are some cases which make no sense at first glance. For instance, the agent is constantly punished when he is not on the road but in the meadow, even if the agent is driving correctly. He will only be rewarded if he drives correctly, is on the road and drives over the randomly placed tiles on the road.

10.5 Results

In this section we discuss the performance of the proposed framework, called Mihaltan2020. The measurement of the performance is done in a typical RL way, namely reward per episode. The training is executed three times with the same hyperparameters and then the median is calculated from the recorded reward per episode. In order to make a deep qualitative investigation of performance, the test bench [11] is used. This consists of four metrics that are derived from the reward-episode curve. The metrics allow a better comparison between frameworks or algorithms than the simple reward-episode curve. The metrics are as follows:

1. Quickness: The amount of episodes that the algorithm needs to arrive at a reward of 700.
2. Stability: The probability of staying over a reward of 700 after achieving that score for the first time.
3. Convergence: The amount of episodes that are consecutively over a reward of 700.
4. Maximum mean reward: The maximum average reward over 100 episodes.

Figure 10.9 shows a comparison between the Acar2019 and Mihaltan2020 Framework. The Acar2019 is an RL framework, which has a half-shared network architecture and trains the entire neural network exclusively via the RL algorithm PPO. Unlike normal PPO frameworks, some of the CNN layers are shared between actor and critic and trained together. In [1] Acar et al. (2019) master thesis it was shown that such an architecture delivers better results than other constellations. The Mihalten2020 architecture is the method that was mentioned in section 10.3. As can be seen in Fig. 10.9, the Mihaltan2020 framework outperform the traditional RL methods.

In Fig. 10.10 the difference between the two methods becomes even clearer with the help of the four metrics. If we look at the means of the metrics, the Mihaltan2020 structure wins on all levels. Especially in terms of stability and convergence the difference is more significant. Furthermore, it is also visible that the variance of Mihaltan2020 is much smaller.

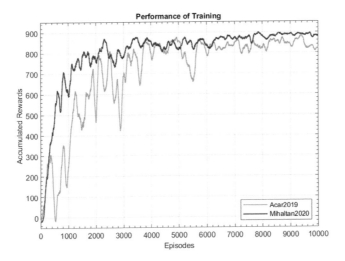

Figure 10.9: Performance comparison between the Acar2019 and the Mihalten2020 Framework. Acar2019 is a state of the art PPO framework, which is used as a comparison to verify the framework according to section 10.3 (Mihaltan2020).

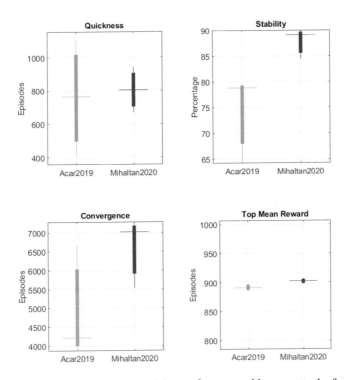

Figure 10.10: Investigation of the training performance with respect to the four metrics.

10.6 Conclusion

The results from section 10.5 show a clear difference between the two investigated methods. It supports the assumption that a modular design of the network architecture is recommended. In addition, the use of different learning rules from different classes than just RL is recommended. Training a perception module via unsupervised learning provides a module that is generally able to decompose inputs into global and general features. It is global because the FE is not dependent on RL and therefore on the task. These are then presented to the RL agent as inputs. Isolating the RL to the actor network ensures stabilization of the whole framework.

Acknowledgment

This work has been funded by the Federal Ministry of Transport and Digital Infrastructure (BMVI) within the funding guideline "Automated and Connected Driving" under the grant number 16AVF2134.

References

1. Acar, K., S.T. Mohammed and G. Ascheid. 2019. Investigation of the PPO algorithm for autonomous driving. ICE RWTH Aachen. URL: https:// www.ice.rwth-aachen. de/.
2. Baidu, P. 2019. Apollo perception module. URL: http://apollo.auto/platform/ perception.html [Online; accessed 7. Jan. 2020].
3. Baldi, P. 2012. Autoencoders, unsupervised learning, and deep architectures. PMLR, pp. 37–49. URL: http://proceedings.mlr.press/v27/baldi12a.html.
4. Geitgey, A. 2018. Machine learning is fun part 6: How to do speech recognition with deep learning. Medium. URL: https://medium.com/@ageitgey/machine-learning-is-fun-part-6-how-to-do-speech-recognition-with-deep-learning-28293c162f7a
5. Goodfellow, I., Y. Bengio and A. Courville. 2016. Deep Learning. MIT Press. http:// www.deeplearningbook.org.
6. Ha, D. and J. Schmidhuber. 2018. World Models. arXiv. doi:10.5281/zenodo.1207631. arXiv:1803.10122.
7. Hoffman, M.D. and D.M. Blei. 2014. Structured stochastic variational inference. arXiv. URL: https://arxiv.org/abs/1404.4114. arXiv:1404.4114.
8. Igl, M., L. Zintgraf, T.A. Le, F. Wood and S. Whiteson. 2018. Deep Variational Reinforcement Learning for POMDPs. arXiv. URL: https://arxiv.org/abs/1806.02426. arXiv:1806.02426.
9. Kingma, D.P. and M. Welling. 2013. Auto-encoding variational bayes. arXiv, URL: https://arxiv.org/abs/1312.6114. arXiv:1312.6114.
10. Krizhevsky, A., I. Sutskever and G.E. Hinton. 2012. Imagenet classification with deep convolutional neural networks. URL: https://papers.nips.cc/paper/4824-imagenet-

classification-with-deep-convolutional-neural-networks [Online; accessed 7. Jan. 2020].

11. Laux, H., S.T. Mohammed and G. Ascheid. 2019. Deep reinforcement learning for autonomous driving with auxiliary tasks. ICE RWTH Aachen. URL: https://www.ice. rwth-aachen.de/.

12. LeCun, Y., B. Boser, J.S. Denker, D. Henderson, R.E. Howard, W. Hubbard and L.D. Jackel. 1989. Backpropagation applied to handwritten zip code recognition. *Neural Comput.*, 1, 541–551. doi:10.1162/neco.1989.1.4.541.

13. LeCun, Y., C. Cortes and C.J. Burges. 2013. Mnist handwritten digit database, yann lecun, corinna cortes and chris burges. URL: http://yann.lecun.com/exdb/mnist [Online; accessed 7. Jan. 2020].

14. Lillicrap, T.P., J.J. Hunt, A. Pritzel, N. Heess, T. Erez, Y. Tassa, D. Silver and D. Wierstra. 2015. Continuous control with deep reinforcement learning. arXiv. URL: https://arxiv.org/abs/1509.02971. rXiv:1509.02971.

15. Loiseau, J.-C.B. 2019. Rosenblatt's perceptron, the very first neural network. Medium. URL: https://towardsdatascience.com/ rosenblatts-perceptron-the-very-first-neural-network-37a3ec09038a.

16. Mihaltan, A., S.T. Mohammed and G. Ascheid. 2020. Investigation of novel network architectures for reinforcement learning algorithms applied to autonomous driving. URL: https://www.ice.rwth-aachen.de/.

17. Mnih, V., A.P. Badia, M. Mirza, A. Graves, T.P. Lillicrap, T. Harley, D. Silver and K. Kavukcuoglu. 2016. Asynchronous methods for deep reinforcement learning. arXiv. URL: https://arxiv.org/abs/1602.01783.arXiv:1602.01783.

18. Mnih, V., K. Kavukcuoglu, D. Silver, A. Graves, I. Antonoglou, D. Wierstra and M. Riedmiller. 2013. Playing atari with deep reinforcement learning. arXiv. URL: https://arxiv.org/abs/1312.5602. arXiv:1312.5602.

19. OpenAI. 2019. Gym: A toolkit for developing and comparing reinforcement learning algorithms. URL: https://gym.openai.com/envs/CarRacing-v0 [Online; accessed 7 Jan. 2020].

20. Radford, A. 2019. Better language models and their implications. OpenAI. URL: https://openai.com/blog/better-language-models.

21. Ramon, Y. 2019. How to start with kaldi and speech recognition. Medium. URL: https://towardsdatascience.com/how-to-start-with-kaldi-and-speech-recognition-a9b7670ffff6.

22. Rocca, J. 2019. Understanding variational autoencoders (vaes). Medium. URL: https://towardsdatascience.com/understanding-variational-autoencoders-vaes-f70510919f73.

23. Schulman, J., F. Wolski, P. Dhariwal, A. Radford and O. Klimov. 2017. Proximal policy optimization algorithms. arXiv. URL: https://arxiv. org/abs/1707.06347. arXiv:1707.06347.

Functional Safety of Deep Learning Techniques in Autonomous Driving Systems

Zheyi Chen, Pu Tian, Weixian Liao* and Wei Yu

Department of Computer and Information Sciences, Towson University, Towson, MD, 21252 USA

11.1 Introduction

Motivated by Google DeepMind demonstrations on Atari games and Go game [38, 53], artificial intelligence has attracted tremendous attention on making machines interact with the environment and gain knowledge. One of the ultimate goals of artificial intelligence is to allow machines to automatically perform all sorts of difficult tasks in different sectors, such as extracting information from complex environments, making real-time decisions and control, to name a few [21, 31, 68]. When it comes to autonomous driving, a self-driving vehicle should have full capability to work properly with every possible situation and provide an effective and safety-guaranteed performance during driving process without human intervention [5].

The Society of Automotive Engineers (SAE), a professional association and standards organization in transportation industries, outlines the six-level description of the vehicle automation, as shown in Fig. 11.1, starting with no automation (Level 0), where human driver has to take full responsibility of controlling steering, throttle and breaking, to full automation (Level 5), where the autonomous driving system takes full control of vehicle without any human intervention [17, 36]. With the level increase in Fig. 11.1, the vehicle itself has the increasing capacity to plan the sequence of driving actions such as steering, throttle and breaking, while interacting with other vehicles in a very complex and latency sensitive environment. The ultimate goal of autonomous driving is to realize that all decisions in the driving process are fully made by a vehicle.

---

*Corresponding author: wliao@towson.edu

	SAE level	Name	Description
The human driver monitors the environment	0	No Automation	Human driver is responsible for steering, throttle and breaking.
	1	Driver Assistance	The vehicle can perform some control function but not everywhere.
	2	Partial Automation	The vehicle can handle steering, throttle and breaking but the driver is expected to monitor the system and take over in case of faults.
The driving system monitors the environment	3	Conditional Automation	The vehicle monitors the surroundings and notifies the driver if manual control is needed.
	4	High Automation	The vehicle is fully autonomous but only in defined use cases
	5	Full Automation	The driver has only to set the destination. The vehicle will handles any surrounding and make any kind of decision on the way.

Figure 11.1: The varying-level description of vehicle automation [36].

Both industry and academia have been making great efforts to envision mature autonomous driving applications on the road by 2021 [69].

To achieve this goal, artificial intelligence has provided a number of powerful techniques as promising solutions to accelerate this process, making vehicles become a rational machine to handle challenging driving tasks automatically. Thanks to outstanding performances realized by deep learning techniques, a number of implementations have been developed and applied in advanced driver assistance systems, aiming to improve the existing functionalities in the vehicle. Examples include Adaptive Cruise Control (ACC), Forward Collision Warning (FCW), Lane-keeping Assist (LKA) and Intelligent Speed Assistance (ISA), among others [1].

Nonetheless, some security/safety-critical applications supported by deep learning techniques, including traffic sign recognition, traffic surveillance, and driver safety assistance, among others, are not fully covered by the existing functional safety standards, such as ISO 26262. ISO 26262, entitled "Road vehicles Functional safety", which is a safety standard in automotive domain used to regulate functional safety and address the safety requirements [28]. As the standard does not catch up with the exploding development of deep learning schemes in vehicle industry, it raises some unknown functional safety risks. For example, the current ISO 26262 standard does not consider the existence of adversarial attackers with malicious intentions that target the vulnerability of deep learning techniques in autonomous driving. Thus, it is of pressing importance to investigate the newly potential functional safety risks of deep learning techniques and their reliability in autonomous driving.

In this chapter, we focus on the existing works in safety risks of deep learning technologies in both general applications and autonomous driving systems. Most deep learning schemes applied in autonomous driving systems can generally be divided into three categories: recognition, prediction and planning. We investigate the existing research on revealing and mitigating potential functional safety risks of deep learning techniques applied in autonomous driving, which could significantly affect safety and performance. The origin of security risks is organized as three aspects: (i) non-transparency, (ii) data and environment uncertainties, and (iii) external adversarial attacks.

Specifically, most of developed deep learning schemes in autonomous driving domain are non-transparent, i.e., being black-box systems, making them extremely difficult for system assessors to develop sufficient confidence that the trained model in design time will operate as intended in run time. Then, deep learning models are trained by a given dataset, which may not be generalized sufficiently to cover varying and complex environments. On the other hand, there is no guarantee that collected training samples can be truly representative of real-world situations such that sufficient coverage can be provided to represent the space of all possible inputs. Thus, the training-based deep learning systems in real-world scenarios often lead challenging issues such as the incompleteness of training data, difference between training and operational environments, and uncertainty of functionalities (e.g., recognition, prediction, and planning). Similarly, the uncertainty of deep learning techniques also raises in the complex real-world circumstances, including execution context, multiple ownership, human involvement, and so on.

Some research efforts have shown that deliberative adversarial attacks can easily fool a well-trained learning model (e.g., traffic sign recognition) and cause catastrophic damages. For example, an attacker could inject some poisoning data in the training phase to prevent the trained model from correctly recognizing traffic signs. Based on the aforementioned issues, we will present the design challenges involved with the safety concerns of deep learning techniques, and investigate the potential solutions to address these new challenges. In addition, we will construct several reasonable examples to show that the aforementioned issues clearly pose potential security risks over different real-world scenarios. Finally, we outline some future research directions in the reliable and fault tolerant implementation of deep learning techniques in highly automated driving.

The remainder of this chapter is organized as follows. In Section 11.2, we survey the existing work over autonomous driving based on three different classes. In Sections 11.3, 11.4 and 11.5, we investigate the existing work on potential security issues of deep learning techniques in aspect of non-transparency, data and environment uncertainties, and external adversarial attacks, respectively. We also present several reasonable examples to illustrate the aforementioned issues and outline some future research direction. Finally, in Section 11.6, we conclude the paper.

11.2 AI and Autonomous Driving

Artificial intelligence is considered a promising solution in the automobile industry which can deliver autonomous driving to the real-world road. In the context of artificial intelligence, technologies such as deep learning have been widely accepted due to their significant capabilities over various tasks. Deep learning, a subset of machine learning in artificial intelligence, is a powerful method to process high-dimensional and complex data based on a hierarchical level of artificial neural networks [7, 50]. The artificial neural networks utilize a web structure, in which neuron nodes are connected together to imitate the human brain. Compared to traditional programs which process data in a linear way, a hierarchical structure makes machines process data in a nonlinear manner [30].

Generally speaking, deep learning models are trained by a huge predefined dataset in a data center environment. Then, trained models are applied to vehicles so that deep learning models can use the information gathered by senors on the vehicle to make decisions and predictions. To be specific, a camera or Lidar sensor collects information from the environment. Then, sensor fusion technologies are used to translate these raw data to their semantic representation. Finally, deep learning models give results by using these preprocessed data in various tasks (e.g., traffic sign recognition, road detection, and others). Recall that there are three main aspects to catalog deep learning technologies over autonomous driving, including recognition, prediction and planning.

Deep learning has been the most popular and powerful tool in recognition domain over the last few years, where it achieves impressive results in a number of well-known computer vision competitions, such as ImageNet [2]. In autonomous driving, vehicles use such learning techniques to identify and recognize the specific object from the surrounding environment, including object detection, pedestrian detection, traffic sign recognition, and road detection, among others. Other functions (friction avoidance, traffic flow prediction, etc.) could use the results from recognition applications to make predictions.

Some existing research efforts have proposed to combine different Convolutional Neural Networks (CNNs) and Support Vector Machines (SVMs) on recognition tasks [59, 60, 66]. To be specific, Uçar et al. [60] proposed a deep learning scheme for pedestrian detection and object recognition, in which CNN and SVM are combined to process the complex real-world environment such as background clutters, the presence of shadow, partial occlusion and the variations in light intensity. Discriminative features in the divided original images were extracted by a nine-layers CNN. Then, the Principal Component Analysis (PCA) was applied to further extract features. Finally, features obtained by PCA were imported to the SVM classifiers. Their experiments were conducted over two datasets, including the Caltech-101 (the per-class accuracy) and Caltech pedestrian dataset (the average miss rate). Self-driving vehicles are considered a model, which makes prediction by using the current state of surrounding environment. Recurrent Neural Network (RNN) can be used to act such as task by keeping sequences of data in their internal state [41]. In addition, CNN can predict steering angle in autonomous driving by the collected raw data from a camera [11].

Combining recognition and prediction can be used to plan the sequence of driving actions in autonomous driving. Such demonstrations have been shown by using a reinforcement learning framework. Generally speaking, reinforcement learning (RL), one branch of machine learning, enables an agent to find a suitable strategy in its action set, so that it can maximize a numerical reward function [56]. For example, Sallab et al. [49] proposed a deep reinforcement learning framework to perform the end-end task in an autonomous driving application, where deep learning algorithms and Recurrent Neural Networks (RNNs) were used for recognition and prediction, respectively. The proposed deep RL framework was validated via the Open-source Racing Car Simulator (Torcs), which demonstrates the correct sequence of driving actions, keeping vehicles in the lane with a good control of the speed limit.

11.3 Non-Transparency Based Safety Risks

Recently, deep learning technologies have achieved great successes on various tasks such as recognition, prediction and planning [7]. Non-transparency refers to most machine learning techniques, including deep learning, are black-box systems for humans because knowledge extracted by encoding is difficult to interpret.

In general, the black-box problem can be defined as a decision-making process provided by learning algorithms is difficult to fully understand. Also, decisions or outputs produced by learning models cannot be correctly predicted [9]. From the machine's view, data features are extracted efficiently and effectively by a multi-layer DNN architecture, but it is difficult for humans to understand these intermediate data. Figure 11.2 shows that data samples (e.g., text messages, voice and images) are brought into a multi-layer black-box system, through which results are computed. The non-transparency characteristic of deep learning techniques contributes to the security risks in autonomous driving applications as it makes the evaluation process much difficult.

Machine learning techniques lacking of transparency could raise issues for intent and causation tests as well. To be specific, intent test is a common method which appears throughout the law and have developed over centuries to help courts and juries understand and regulate human behaviors. Nonetheless, when it comes to autonomous driving, non-transparency brings an obstacle to find out whether a person (designer) or computer programs itself intended to cause a particular outcome occur in an accident [6].

To address aforementioned concerns, some studies categorize the non-transparency issues to the following two types: (i) *Prediction interpretation (explanation) and justification*: the explanation of a prediction result over a black-box system must be generated; (ii) *Interpretable models*: Aiming at designing models that are intrinsically interpretable [9, 22, 40, 52]. Some studies show that explanation and justification generally improve user satisfaction and acceptance of the developed system, as well as make the user change their attitudes towards the system [10, 23, 57]. This is important because new applications such as autonomous driving systems need convincing theory as foundations to provide user confidence instead of bringing a black-box to users.

A number of works explain the prediction results of machine learning models by considering the impacts of features independently [29, 47]. For example, Robnik-Šikonja et al. [47] studied the explanation of decisions made by a probabilistic radial basis function (PRBF) network, which is an artificial neural network (black-box system). A decomposition of PRBF network was proposed by utilizing the marginalization property of the Gaussian distribution, which translates the PRBF network to a white-box non-linear system. Likewise, a demonstration was shown that approaches in [46, 55] could successfully explain the results of unseen data samples in PRBF network as well as a graphical description of non-transparency model was delivered. Particularly, Poulin et al. [42] focused on the case of additive classifiers; however, the extracted features in high level cannot be directly translated to an additive function in the low level network. Landecker et al. [29] developed an approach, namely contribution propagation, which not only provides the explanation for hierarchical networks (a general class in deep learning), but also addressed the aforementioned challenge. To be specific, the authors studied the importance of each feature with a given data sample. Note that the contributions of each high level feature computed by the method proposed in [42] are related backward to the low level network. The relative contribution between the node in the low level network and their parent

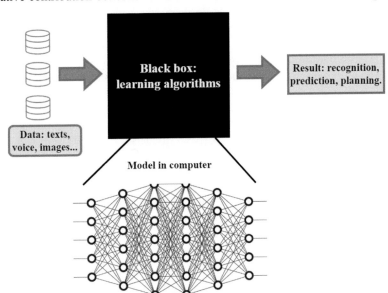

Figure 11.2: The black-box systems in machine learning and deep learning.

nodes are also determined. Thus, contribution propagation scheme produces the explanation of individual classifications. The effectiveness of the proposed scheme in [29] was validated via two experiments, in which one is an artificial visual classification over simple shapes and the other is conducted over a real-world dataset (Caltech101 dataset).

Nomogram, allowing that a mathematical function can be approximately represented by a graphical scheme, is another common explanation technique for machine learning. For example, Belle et al. [62] studied the explanation of Support Vector Machine (SVM) with a color based nomogram. To explain SVM with linear, polynomial and RBF kernels, the authors approximated the SVM model by a sum of main and two-way interaction terms, where nomogram yields a visual explanation. Moreover, it shows that lines in nomogram are replaced by colors bars as well as kernels, in which kernel parameters can be exactly represented by the proposed color based nomogram. In addition, Ribeiro et al. [45] studied trusting a prediction and a model in the machine learning. An algorithm, called LIME, was proposed to yield the explanation of prediction results that first aims to solve the problem on trusting a prediction. Then, combining multiple prediction results provided by LIME solves the problem of trusting a model. To be specific, LIME is capable of approximating a prediction result produced by using an interpretable model. The aim of LIME is to identify an interpretable model over the interpretable data representation, instead of features used by machine learning model.

Some other works make efforts on interpretable models to eliminate non-transparency in various learning models. For example, Ba and Caruana [4] raised a few interesting questions, such as do deep nets really need to be deep and what is the source of this improvement? To answer these questions, the schemes against non-transparency need to further be studied, at least in empiricism. Che et al. [12] introduced a scheme, namely interpretable mimic learning, which was designed to extract knowledge and learn interpretable models as deep learning models with robust prediction results. Instead of the original interpretable mimic learning, exploiting shallow neural networks or kernel methods, gradient boosting trees [12, 18] were used to learn interpretable models from Deep Neural Network (DNN) and Recurrent Neural Network (RNN) models. Likewise, Ustun et al. [61] created a sparse model by features selection over data-driven scoring systems. The proposed scheme aims to produce interpretability, which is achieved by optimizing the accuracy and sparsity. Similar to [61], Kim et al. [27] proposed a scheme for interpretable model by means of feature extraction and selection. In addition, the proposed scheme adds interpretable criteria into the mode, rather than using certain distance metric to measure the variation of the predictions and post-processing to provide interpretable models.

All aforementioned research efforts have achieved exciting results on revealing non-transparency. To combine the different learning techniques with autonomous driving, non-transparency problem still needs further research. In some cases, Fig. 11.3 shows that one self-driving vehicle that carries someone

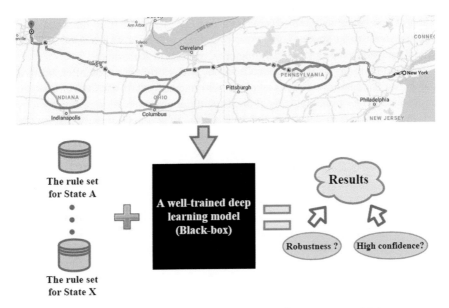

Figure 11.3: Updating non-transparency system in autonomous driving.

to pass through a couple of states, but each state may have different regulations and polices. These cases clearly raise a few unsolved questions, including (i) *Can we add these new regulations and polices (rules) in a trained model directly?* (ii) *How can we achieve this model extension for such a black-box system?* (iii) *Does it pose new potential security risks?*

Non-transparency leads to robustness problem which raises security risks in all autonomous applications, including autonomous driving. For example, we wonder whether a self-driving vehicle can operate correctly over invalid inputs (an anomaly traffic sign) or stressful environmental conditions (extreme weather conditions). Such problems have raised significant challenge for all other artificial intelligence applications. This is because we live in a real world, where most things work in rule-based manners, but most machine learning and deep learning models are data-driven systems, which cannot be translated to a rule-based manner directly. In the worst case, we can train a new learning model to solve such problems, but training a new machine learning model is expensive and time-consuming.

11.4 Data and Environment Uncertainties in Autonomous Driving Systems

Data incompleteness and a complex environment bring uncertainty to the learning systems [37, 51, 64]. Most machine learning and deep learning (ML/DL) models are trained with a predefined training dataset by minimizing the loss function.

There is no guarantee to the selected data samples, which are truly representative over the real-world space. Thus, error rate generated by a well-trained ML/DL model may not represent its performance in real-world scenarios.

In autonomous driving applications, the real-world scenarios often bring a number of uncertainties due to complex environment, yet self-driving vehicles are envisioned to be a predictable machine. One important application in autonomous driving is traffic sign detection and recognition, which may directly lead to the self-driving vehicles make the different sequence of driving actions. Some works make efforts on this area [25, 43, 70] to provided effective performance. For example, Qian et al. [43] proposed a convolutional neural networks (CNN) based scheme to conduct feature extraction and classification. The effectiveness of the proposed scheme was evaluated on several popular datasets (GTSRB, MNIST, CASIA GB1, LP Dataset and GTSDB), achieving impressive results. The research efforts similar to [43] leverage the outstanding capability of CNN/DNN over image classification, localization and detection to achieve impressive on their performance metrics, including recall and accuracy (over 95% on average).

These developed ML/DL models achieve good performance on public standard datasets, showing that machine learning and deep learning techniques have the strong capability on both data feature extractions and classification tasks. However, all ML/DL models developed by aforementioned works have a common issue that is the incompleteness of both training datasets and testing datasets. Data incompleteness brings potential security risks to autonomous driving. Imagine that a self-driving vehicle is running in an urban environment where the road speed limit is only 25 mph (school zone in U.S), but that speed limit sign is partially blocked by some obstacles. As shown in Fig. 11.4, these cases are common to appearing in people's daily lives. The inappropriate sequence of driving actions raise the security risks, because the ML/DL models cannot provide a correct recognition.

The most datasets mainly represent an ideal environment, which assumes data features can be 100% exposed to the learning process. The gap between the ideal data-center environment and real-world environment is not eliminated, even if some works [25, 44, 70] created a number of more realistic traffic-sign benchmarks to validate the effectiveness of ML/DL models. This is because the collected data samples may not represent the safety-critical cases. To bring the mature autonomous driving on the road, behaviors of ML/DL applications under aforementioned events still need further investigation. Except partial occlusion, the influence of partial occlusion, the presence of shadows, the variations in light intensity, and the different weather conditions should be taken into the consideration. It is clear that a complete dataset for a real-world case becomes extremely large, if we add all above-mentioned variables (partial occlusion, the presence of shadows, different weather conditions, and others) to any existing datasets.

Another problem is noise in the dataset, even if we have already built a nearly perfect dataset. Prior research shows that deep neural networks could achieve

Figure 11.4: The partial occlusion of traffic signs in real world.

impressive performance over a noisy dataset [48, 63]. To be specific, noisy data samples can be generated by randomly mapping the training samples with wrong labels. It is no doubt that these works illustrate learning processes can be robust to a given amount of label noise, if clean data is sufficiently provided to that learning process. Nonetheless, can anyone leverage this powerful capability of learning model to achieve their hidden expectation? Unfortunately, a number of research efforts showed that noise pattern and powerful data extraction capability could raise potential security issues in autonomous applications [8, 24]. We will discuss these works in next section in detail.

Uncertainty, in an unpredictable and stochastic environment, is everywhere, because autonomous driving is requested to engage in all sorts of situations where customers give the destination for his/her trips without any other driving actions [36]. In fact, the problem is that we cannot make the pre-defined data samples to fully represent all possible scenarios. For example, some extreme weather conditions cannot be precisely foreseen so far. Thus, to deal with uncertainty, the proposed model shall have the capability to evolve itself and adapt to an unforeseeable environment [35, 65].

Recently, the research topic, namely stochastic programming, plays an critical role to address uncertainty problems in self-adaptive software systems [26]. Nonetheless, stochastic programming is a resource-intensive technique, implying a computationally expensive and data intensive process. These characteristics makes it hard to operate in real-time applications, where autonomous applications have to make decisions rapidly. Additionally, a number of works have been devoted to quantifying uncertainty [14, 15]. For instance, Esfahani et al. [14] showed that a self-adaptive system could achieve better results than previous works even if uncertainty in the system is partially quantified. Such works are challenging as the source of uncertainty is ubiquitous in all sorts of environments.

Overall, a self-adaptive system framework allows self-driving vehicles to gather the real-world information from their operation scenarios, update their system states by importing these fresh data, and then adapt to events which are never seen in original datasets. Such self-adaptive techniques also raise new challenges for both academia and industry, because it is similar to manage a large-scale distributed system, which requests a full self-awareness on itself [32].

11.5 External Adversarial Attacks

Autonomous driving cars rely on real-time data collected by in-car cameras, speed radar system, motion detectors, LiDAR light sensors, and the GPS location system for the DNN model to make driving decisions. Nonetheless, DNN models are vulnerable to elaborately crafted adversarial examples. For self-driving cars, wrong decisions could pose serious security concerns when compromised input data are fed into the DNN model: (i) Adversarial traffic signs could deceive the system to misclassify a red light as green; (ii) Noisy radio interference could malfunction GPS/speed sensors. When self-driving cars are moving at high speed, adversarial attacks may result in fatal accidents. In the following, we discuss camera related attacks and defenses in detail.

Adversarial attack is first discussed in [3, 33, 58]. In their study, authors claimed that an image could be classified incorrectly using DNNs by maximizing the prediction error through adding imperceptible perturbations. For example, Fig. 11.5 in [19] illustrates that the prediction result could be misled by adding imperceptible noises. For autonomous driving, added noises, with the weight factor $\beta(0 < \beta < 1)$, to a traffic sign sample could lead to wrong prediction result, similar to one shown in Fig. 11.6.

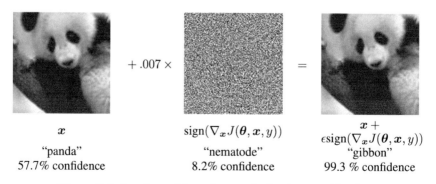

$$+ .007 \times \qquad =$$

$$x \qquad \qquad \text{sign}(\nabla_x J(\boldsymbol{\theta}, \boldsymbol{x}, y)) \qquad \begin{array}{c} x + \\ \epsilon\text{sign}(\nabla_x J(\boldsymbol{\theta}, \boldsymbol{x}, y)) \end{array}$$

"panda"   "nematode"   "gibbon"
57.7% confidence   8.2% confidence   99.3 % confidence

Figure 11.5: An adversarial attack example: Panda recognition [19].

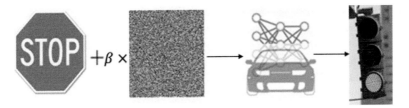

Figure 11.6: Traffic sign adversarial attack.

Traffic sign recognition is essential for auto driving. There are some research efforts in this direction. For example, Sitawarin et al. [54] investigated the Lenticular attack, which exploits the angle difference of driver view and the camera view by integrating adversarial examples into a single sign. Likewise, Eykholt et al. [16] extended the study of physical attack to object detection models. The authors improved the RP-2 algorithm to launch solid object detection attacks, aiming to mislead classifiers by either covering the sign or adding noises using adversarial stickers. In their experiments, they successfully invalidated a model to recognize the existence of a STOP sign and they also malfunctioned one of the state-of-art models YOLO v2 with the misclassification rate up to 85% in lab environment and 63.5% in real environment. In addition, authors also validated the attack transferability of their attacks using R-CNN object detector. In the result, their designed scheme could fool the R-CNN model with 85.9% in the lab and 40.2% in a real outdoor environment.

Some works focus on poisoning attacks also achieve a few remarkable results [8, 13]. To be specific, poisoning attacks refers to the one that attackers inject some poisoning data samples into the training process to achieve their desired goals. For example, Chen et al. [13] studied the targeted poisoning attack against a deep learning model. Their results show that it is feasible that an attacker could add a desired backdoor in deep learning models by injecting a small number of dirty data samples. In the testing phase, these backdoors allow the attacker easily use some elaborate samples with a physical key to make the trained models misclassify to a desired result. Some potential detection and mitigation strategies were tested. However, none of these strategies could detect or mitigate either dirty data samples or targeted backdoor over the learning system. Thus, poisoning attacks in the deep learning system raise the new security issues in autonomous systems. For example, when an attacker injects such a backdoor, the result of a traffic-sign recognition system in autonomous driving can be manipulated. Likewise, in [44], authors studied the robustness of optical flow under adversarial attacks. They showed that attacks with corrupted patches could impact the performance significantly. They also pointed out the extent of negative influences might vary with different DNNs where encoder-decoder networks suffer more compared with pyramid networks. In the end, they proposed a tool for attack detection for the flow network to improve the robustness.

After reviewing some attack strategies, we now discuss some defense strategies for adversarial attacks in two aspects: (i) *training data modification* and (ii) *model modification*. To the best of our knowledge, as time of this writing, we could not find any academic research specific in autonomous driving adversarial defense. Thus, we list some typical defense strategies and discuss how they can be used to protect autonomous driving.

- **Modification defense.** This defense strategy tends to conduct data modification. The raw training data is first modified before being fed into the DNN model. Related to this strategy, Xie et al. [67] proposed a randomization scheme to mitigate adversarial attacks. For data instances, they are first modified with two randomization operations. One is the random resizing that changes the size into a random state and the other is the random padding that inserts zeros around the resized image. In their experiments, they claimed that their scheme works effectively for both single-step and iteration attacks. In addition, they argued that the proposed scheme does not need extra training steps for the DNN, and can be used in combination with other defense strategies. Sample reconstruction is another defensive strategy. Related to this strategy, Gu and Rigazio [20] designed a denoising autoencoder to clean adversarially added perturbations.

- **Model modification.** This defense strategy makes changes to the DNN model structure. For instance, Liu et al. [34] proposed the Random Self-Ensemble (RSE) scheme to prevent gradient attacks by adding noise layers randomly in the DNN model. In each noisy layer, some tiny noises are added to the data from the previous layer. By inserting noises, the designed scheme aims to eliminate the attacker's perturbation noises. Likewise, network distillation [39] uses the first model's probability distribution vector output as the original model's input to another DNN. Such a process can counteract perturbations' negative impacts and thus increases the network's robustness to adversarially compromised inputs.

By looking into the investigated attack and defense strategies, we conclude that DNN based driving techniques are challenging with adversarial attacks in toxic traffic signs, poisonous training data, and compromised sensors. Although we identify some defense strategies designed for dealing with general adversarial attacks, there is still no systematic defense strategies which are capable of dealing with the security issues in autonomous cars. Thus, by reviewing existing defense strategies, we give our suggestions for self-driving systems so that the security risk can be reduced.

- **Pre-knowledge.** Autonomous driving systems can combine with other information sources such as GPS, street view to obtain more specific data to the road. In addition, they can upload their captured data to the data center to help spot potential attacks.

- **Self-adaptive learning.** In some cases, drivers may take some actions to rectify wrong decisions of the self-driving system. Those actions can be used to help retrain the system to adapt driving environment changes.

11.6 Conclusion

Autonomous driving has attracted increasing attentions in both academia and industry. Artificial intelligence has been considered as a promising solution, which can significantly support fully autonomous driving in the real world. It shows that deep learning has achieved impressive performances, including recognition, prediction and planning. In this chapter, we have surveyed different deep learning techniques, a subset of artificial intelligence, over various autonomous driving applications. We notice the current safety standard, such as ISO 26262, cannot well address safety requirements caused by deep learning techniques. We therefore investigate the potential security risk from three different aspects: model non-transparency, uncertainty, and external adversarial attacks. We also construct several reasonable examples to demonstrate the aforementioned issues and point out some future research directions in the reliable and fault tolerant implementation of deep learning techniques in highly automated driving systems.

Acknowledgements

The work was supported in part by the US National Science Foundation (NSF) under grants: CNS 1350145. Any opinions, findings and conclusions or recommendations expressed in this material are those of the authors and do not necessarily reflect the views of the funding agency.

References

1. 40+ corporations working on autonomous vehicles. https://www.cbinsights.com/research/autonomous-driverless-vehicles-corporations-list/.
2. Imagenet. http://www.image-net.org/.
3. Arpit, D., S. Jastrzebski, N. Ballas, D. Krueger, E. Bengio, M.S. Kanwal, T. Maharaj, A. Fischer, A. Courville, Y. Bengio, et al. 2017. A closer look at memorization in deep networks. *In*: Proceedings of the 34th International Conference on Machine Learning, 70, 233–242. JMLR. org,
4. Ba, J. and R. Caruana. 2014. Do deep nets really need to be deep? *Advances in Neural Information Processing Systems*, 2654–2662.
5. Badue, C., R. Guidolini, R.V. Carneiro, P. Azevedo, V.B. Cardoso, A. Forechi, L. Jesus, R. Berriel, T. Paixo, F. Mutz, L. Veronese, T. Oliveira-Santos and A.F.D. Souza. 2019. Self-driving Cars: A Survey.
6. Bathaee, Y. 2017. The artificial intelligence black box and the failure of intent and causation. *Harv. JL & Tech.*, 31, 889.

7. Bengio, Y., A. Courville and P. Vincent. 2013. Representation learning: A review and new perspectives. *IEEE Transactions on pattern Analysis and Machine Intelligence*, 35(8), 1798–1828.

8. Bhagoji, A.N., S. Chakraborty, P. Mittal and S. Calo. 2018. Analyzing federated learning through an adversarial lens. *arXiv preprint arXiv:1811.12470*.

9. Biran, O. and C. Cotton. 2017. Explanation and justification in machine learning: A survey. *IJCAI-17 Workshop on Explainable AI (XAI)*, 8, 1.

10. Biran, O. and K.R. McKeown. 2017. Human-centric justification of machine learning predictions. *IJCAI*, 1461–1467.

11. Bojarski, M., D. Del Testa, D. Dworakowski, B. Firner, B. Flepp, P. Goyal, L.D. Jackel, M. Monfort, U. Muller, J. Zhang, et al. 2016. End to end learning for self-driving cars. *arXiv preprint arXiv:1604.07316*.

12. Che, Z., S. Purushotham, R. Khemani and Y. Liu. 2016. Interpretable deep models for ICU outcome prediction. *In*: AMIA Annual Symposium Proceedings, vol. 2016, 371. American Medical Informatics Association.

13. Chen, X., C. Liu, B. Li, K. Lu and D. Song. 2017. Targeted backdoor attacks on deep learning systems using data poisoning. *arXiv preprint arXiv:1712.05526*.

14. Esfahani, N., E. Kouroshfar and S. Malek. 2011. Taming uncertainty in self-adaptive software. pp. 234–244. *In*: Proceedings of the 19th ACM SIGSOFT Symposium and the13th European Conference on Foundations of Software Engineering, ACM.

15. Esfahani, N. and S. Malek. 2013. Uncertainty in self-adaptive software systems. *Software Engineering for Self-Adaptive Systems II*, 214–238. Springer.

16. Eykholt, K., I. Evtimov, E. Fernandes, B. Li, A. Rahmati, F. Tramer, A. Prakash, T. Kohno and D. Song. 2018. Physical adversarial examples for object detectors. *arXiv preprint arXiv:1807.07769*.

17. Favarò, F., S. Eurich and N. Nader. 2018. Autonomous vehicles disengagements: Trends, triggers, and regulatory limitations. *Accident Analysis & Prevention*, 110, 136–148.

18. J.H. Friedman. 2016. Greedy function approximation: A gradient boosting machine. 1999. DOI= http://wwwstat.stanford.edu/jhf/ftp/trebst. pdf.

19. Goodfellow, I.J., J. Shlens and C. Szegedy. 2014. Explaining and harnessing adversarial examples. *arXiv preprint arXiv:1412.6572*.

20. Gu, S. and L. Rigazio. 2014. Towards deep neural network architectures robust to adversarial examples. *arXiv preprint arXiv:1412.5068*.

21. Hatcher, W.G. and W. Yu. 2018. A survey of deep learning: Platforms, applications and emerging research trends. *IEEE Access*, 6, 24411–24432.

22. Hendricks, L.A., Z. Akata, M. Rohrbach, J. Donahue, B. Schiele and T. Darrell. 2016. Generating visual explanations. *European Conference on Computer Vision*, 3–19. Springer.

23. Herlocker, J.L., J.A. Konstan and J. Riedl. 2000. Explaining collaborative filtering recommendations. *In*: Proceedings of the 2000 ACM Conference on Computer Supported Cooperative Work, 241–250. ACM.

24. Jagielski, M., A. Oprea, B. Biggio, C. Liu, C. Nita-Rotaru and B. Li. 2018. Manipulating machine learning: Poisoning attacks and countermeasures for regression learning. pp. 19–35. *In*: 2018 IEEE Symposium on Security and Privacy (SP), IEEE.

25. Jung, S., U. Lee, J. Jung and D.H. Shim. 2016. Real-time traffic sign recognition system with deep convolutional neural network. pp. 31–34. *In*: 2016 13th International Conference on Ubiquitous Robots and Ambient Intelligence (URAI), IEEE.

26. P. Kall, S. W. Wallace, and P. Kall. 1994. Stochastic Programming. Springer.

27. Kim, B., J.A. Shah and F. Doshi-Velez. 2015. Mind the gap: A generative approach to interpretable feature selection and extraction. *Advances in Neural Information Processing Systems*, pages 2260–2268.

28. Kucharski, M., A. Trujillo, C. Dunlop and B. Ahdab. 2012. ISO 26262 software compliance: Achieving functional safety in the automotive industry. Technical Report.

29. Landecker, W., M.D. Thomure, L.M. Bettencourt, M. Mitchell, G.T. Kenyon and S.P. Brumby. 2013. Interpreting individual classifications of hierarchical networks. pp. 32–38. *In*: 2013 IEEE Symposium on Computational Intelligence and Data Mining (CIDM), IEEE.

30. LeCun, Y., Y. Bengio and G. Hinton. 2015. Deep learning. *Nature*, 521(7553), 436.

31. Liang, F., W.G. Hatcher, W. Liao, W. Gao and W. Yu. 2019. Machine learning for security and the internet of things: The good, the bad, and the ugly. *IEEE Access*, 7, 158126–158147.

32. Liao, W., Y. Guo, X. Chen and P. Li. 2018. A unified unsupervised gaussian mixture variational autoencoder for high dimensional outlier detection. pp. 1208–1217. *In*: 2018 IEEE International Conference on Big Data (Big Data), IEEE.

33. Liao, W., S. Salinas, M. Li, P. Li and K.A. Loparo. 2017. Cascading failure attacks in the power system: A stochastic game perspective. *IEEE Internet of Things Journal*, 4(6), 2247–2259.

34. Liu, X., M. Cheng, H. Zhang and C.-J. Hsieh. 2018. Towards robust neural networks via random self-ensemble. pp. 369–385. *In*: Proceedings of the European Conference on Computer Vision (ECCV).

35. Mallozzi. P. 2017. Combining machine-learning with invariants assurance techniques for autonomous systems. pp. 485–486. *In*: Proceedings of the 39th International Conference on Software Engineering Companion. IEEE Press.

36. Mallozzi, P., P. Pelliccione, A. Knauss, C. Berger and N. Mohammadiha. 2019. Autonomous vehicles: State of the art, future trends, and challenges. *Automotive Systems and Software Engineering: State of the Art and Future Trends*, 347.

37. McAllister, R., Y. Gal, A. Kendall, M. Van Der Wilk, A. Shah, R. Cipolla and A.V. Weller. 2017. Concrete problems for autonomous vehicle safety: Advantages of bayesian deep learning. International Joint Conferences on Artificial Intelligence, Inc.

38. Mnih, V., K. Kavukcuoglu, D. Silver, A. Graves, I. Antonoglou, D. Wierstra and M.A. Riedmiller. 2013. Playing atari with deep reinforcement learning. *arXiv*, abs/1312.5602.

39. Papernot, N., P.D. McDaniel, X. Wu, S. Jha and A. Swami. 2015. Distillation as a defense to adversarial perturbations against deep neural networks. corr abs/1511.04508 (2015). *In*: 37th IEEE Symposium on Security and Privacy.

40. Park, D.H., L.A. Hendricks, Z. Akata, B. Schiele, T. Darrell and M. Rohrbach. 2016. Attentive explanations: Justifying decisions and pointing to the evidence. *arXiv preprint arXiv:1612.04757.*

41. Pinheiro, P.H. and R. Collobert. 2014. Recurrent convolutional neural networks for scene labeling. *In*: 31st International Conference on Machine Learning (ICML), number CONF.

42. Poulin, B., R. Eisner, D. Szafron, P. Lu, R. Greiner, D.S. Wishart, A. Fyshe, B. Pearcy, C. MacDonell and J. Anvik. 1999, 2006. Visual explanation of evidence with additive classifiers. *In*: Proceedings of the National Conference on Artificial Intelligence, 21, 1822. Menlo Park, CA; Cambridge, MA; London; AAAI Press; MIT Press.

43. Qian, R., B. Zhang, Y. Yue, Z. Wang and F. Coenen. 2015. Robust Chinese traffic sign detection and recognition with deep convolutional neural network. pp. 791–796. *In*: 2015 11th International Conference on Natural Computation (ICNC). IEEE.

44. Ranjan, A., J. Janai, A. Geiger and M.J. Black. 2019. Attacking optical flow. *In*: International Conference on Computer Vision (ICCV).

45. Ribeiro, M.T., S. Singh and C. Guestrin. 2016. Why should I trust you? Explaining the predictions of any classifier. pp. 1135–1144. *In*: Proceedings of the 22nd ACM SIGKDD International Conference on Knowledge Discovery and Data Mining. ACM.

46. Robnik-Sˇikonja, M. and I. Kononenko. 2008. Explaining classifications for individual instances. *IEEE Transactions on Knowledge and Data Engineering*, 20(5), 589–600.

47. Robnik-Šikonja, M., A. Likas, C. Constantinopoulos, I. Kononenko and E. Štrumbelj. 2011. Efficiently explaining decisions of probabilistic rbf classification networks. pp. 169–179. *In*: International Conference on Adaptive and Natural Computing Algorithms. Springer,

48. Rolnick, D., A. Veit, S. Belongie and N. Shavit. 2017. Deep learning is robust to massive label noise. *arXiv preprint arXiv:1705.10694*.

49. Sallab, A.E., M. Abdou, E. Perot and S. Yogamani. 2017. Deep reinforcement learning framework for autonomous driving. *Electronic Imaging*, 2017(19), 70–76.

50. Schmidhuber, J. 2015. Deep learning in neural networks: An overview. *Neural Networks*, 61, 85–117.

51. Shafaei, S., S. Kugele, M.H. Osman and A. Knoll. 2018. Uncertainty in machine learning: A safety perspective on autonomous driving. pp. 458–464. *In*: International Conference on Computer Safety, Reliability, and Security. Springer.

52. Shwartz-Ziv, R. and N. Tishby. 2017. Opening the black box of deep neural networks via information. *arXiv preprint arXiv:1703.00810*.

53. Silver, D., J. Schrittwieser, K. Simonyan, I. Antonoglou, A. Huang, A. Guez, T. Hubert, L.R. Baker, M. Lai, A. Bolton, Y. Chen, T.P. Lillicrap, F. Hui, L. Sifre, G. van den Driessche, T. Graepel and D. Hassabis. 2017. Mastering the game of go without human knowledge. *Nature*, 550, 354–359.

54. Sitawarin, C., A.N. Bhagoji, A. Mosenia, M. Chiang and P. Mittal. 2018. Darts: Deceiving autonomous cars with toxic signs. *arXiv preprint arXiv:1802.06430*.

55. Strumbelj, E., I. Kononenko and M.R. Šikonja. 2009. Explaining instance classifications with interactions of subsets of feature values. *Data & Knowledge Engineering*, 68(10), 886–904.

56. Sutton, R.S. and A.G. Barto. 2018. *Reinforcement Learning: An introduction*. MIT Press.

57. Symeonidis, P., A. Nanopoulos and Y. Manolopoulos. 2009. Moviexplain: A recommender system with explanations. *RecSys*, 9, 317–320.

58. Szegedy, C., W. Zaremba, I. Sutskever, J. Bruna, D. Erhan, I. Goodfellow and R. Fergus. 2013. Intriguing properties of neural networks. *arXiv preprint arXiv:1312.6199*.

59. Tao, Q.-Q., S. Zhan, X.-H. Li and T. Kurihara. 2016. Robust face detection using local cnn and svm based on kernel combination. *Neurocomputing*, 211, 98–105.

60. Uçar, A., Y. Demir and C. Güzelis¸. 2017. Object recognition and detection with deep learning for autonomous driving applications. *Simulation*, 93(9), 759–769.

61. Ustun, B. and C. Rudin. 2016. Supersparse linear integer models for optimized medical scoring systems. *Machine Learning*, 102(3), 349–331.

62. Van Belle, V., B. Van Calster, S. Van Huffel, J.A. Suykens and P. Lisboa. 2016. Explaining support vector machines: A color based nomogram. *PloS One*, 11(10), e0164568.

63. Van Horn, G., O. Mac Aodha, Y. Song, Y. Cui, C. Sun, A. Shepard, H. Adam, P. Perona and S. Belongie. 2018. The inaturalist species classification and detection dataset. pp. 8769–8778. *In*: Proceedings of the IEEE Conference on Computer Vision and Pattern Recognition.

64 Varshney, K.R. 2016. Engineering safety in machine learning. pp. 1–5. *In*: 2016 Information Theory and Applications Workshop (ITA). IEEE.

65. Weyns, D. 2019. Software engineering of self-adaptive systems. *Handbook of Software Engineering*, 399–443. Springer.

66. Wu, H., Q. Huang, D. Wang and L. Gao. 2018. A cnn-svm combined model for pattern recognition of knee motion using mechanomyography signals. *Journal of Electromyography and Kinesiology*, 42, 136–142.

67. Xie, C., J. Wang, Z. Zhang, Z. Ren and A. Yuille. 2017. Mitigating adversarial effects through randomization. *arXiv preprint arXiv:1711.01991*.

68. Xu, H., W. Yu, D. Griffith and N. Golmie. 2018. A survey on industrial internet of things: A cyber-physical systems perspective. *IEEE Access*, 6, 78238–78259.

69. Yaqoob, I., L.U. Khan, S.A. Kazmi, M. Imran, N. Guizani and C.S. Hong. 2019. Autonomous driving cars in smart cities: Recent advances, requirements, and challenges. *IEEE Network*.

70. Zhu, Z., D. Liang, S. Zhang, X. Huang, B. Li and S. Hu. 2016. Traffic-sign detection and classification in the wild. pp. 2110–2118. *In*: Proceedings of the IEEE Conference on Computer Vision and Pattern Recognition.

Index